T0296495

LONDON MATHEMATICAL SOCIETY LECTURE NOTE SERIES

Managing Editor: Professor Endre Süli, Mathematical Institute, University of Oxford,
Woodstock Road, Oxford OX2 6GG, United Kingdom

The titles below are available from booksellers, or from Cambridge University Press at
www.cambridge.org/mathematics

London Mathematical Society Lecture Note Series: 477

Elliptic Regularity Theory by Approximation Methods

EDGARD A. PIMENTEL
University of Coimbra

CAMBRIDGE
UNIVERSITY PRESS

CAMBRIDGE
UNIVERSITY PRESS

Shaftesbury Road, Cambridge CB2 8EA, United Kingdom

One Liberty Plaza, 20th Floor, New York, NY 10006, USA

477 Williamstown Road, Port Melbourne, VIC 3207, Australia

314–321, 3rd Floor, Plot 3, Splendor Forum, Jasola District Centre, New Delhi – 110025, India

103 Penang Road, #05–06/07, Visioncrest Commercial, Singapore 238467

Cambridge University Press is part of Cambridge University Press & Assessment,
a department of the University of Cambridge.

We share the University's mission to contribute to society through the pursuit of
education, learning and research at the highest international levels of excellence.

www.cambridge.org
Information on this title: www.cambridge.org/9781009096669

DOI: 10.1017/9781009099899

First published 2022

A catalogue record for this publication is available from the British Library

ISBN   978-1-009-09666-9   Paperback

לאמיליא
שאני באמת אוהב בכל ליבי.
ולמי זה יכול להיות מספיק.

(To Emilia
Who I truly love with all my heart.
And to whom it might suffice.)

Contents

Preface

This set of notes focuses on regularity theory for elliptic partial differential equations (PDEs). In particular, it details regularity results obtained through approximation (perturbative) methods, in line with the techniques launched in the works of Caffarelli (1988, 1989).

Our goal is to tell a story; it starts with the fundamental breakthroughs of the 1980s, namely, the Krylov–Safonov and the Evans–Krylov results for nonvariational PDEs, jointly with Caffarelli's regularity theory for fully nonlinear elliptic equations. At this point we emphasize the importance of convexity. While uniform ellipticity implies Hölder continuity of the gradient, a convexity condition leads to $C^{2,\alpha}$-regularity. Our perspective is that such a fact entails two fundamental directions. First, one asks whether or not $C^{1,\alpha}$-regularity is optimal for merely elliptic equations. Then, one seeks conditions, weaker than convexity, that are capable of unlocking *general* regularity results in between the Krylov–Safonov and the Evans–Krylov theories.

The first direction is well understood and documented in the corpus of results due to Nadirashvili and Vlăduţ (2007, 2008, 2011). As far as the second direction is concerned, an important bifurcation arises. On one hand, differentiability of the operator stands out as a condition affecting the regularity of the solutions.

In this regard, we discuss the developments due to Savin (2007), namely, the fact that flat solutions to equations driven by operators of class C^2 are locally of class $C^{2,\alpha}$. It is worth noticing the contribution in Savin (2007) transcends this discussion as it relates also to the case of degenerate elliptic equations. A further development under differentiability conditions is the so-called partial regularity result, due to Armstrong et al. (2012).

The alternative route concerns the analysis of particular models through a set of techniques capable of producing new information on the regularity of the problems under analysis. We finish the story by detailing the use of approximation and perturbative techniques in the context of a few examples.

We open the book with the basics in the theory of elliptic, discussing elementary notions and results and detailing preliminary facts. This is followed by an account of paramount results, from the Krylov–Safonov theory to the counterexamples of Nadirashvili–Vlăduţ. Chapter 2 is differentiability land; we present the $C^{2,\alpha}$-regularity theory for flat solutions and detail the partial regularity result. In Chapter 3 we introduce the notion of the recession operator and explore its consequences on the regularity of viscosity solutions in Sobolev and Hölder spaces. Chapter 4 examines an important example, namely, the Isaacs equation. The relevance of this model follows from its lack of convexity and differentiability. Here, approximation methods relate that operator to a Bellman one. As a consequence, we establish regularity results for L^p-viscosity solutions. We close the book in Chapter 5 with the analysis of degenerate problems, including a fully nonlinear model as well as a perturbative analysis of the p-Poisson equation.

The choice of topics in these notes reflects only *one* possible perspective on the material. There are plenty of models and examples whose regularity theory has been advanced through approximation methods. In particular, we have not included any discussion on nonlocal equations, or on the spillovers of those methods on free boundary problems. The bibliographical notes closing each chapter aim at adding context to the material; once again the choice here is (un)fairly biased.

Usually, people write books because they know things. This is not the case with these notes – quite the contrary. In 2017, during a talk of mine, Professor Diego Moreira asked me a question I was not sure how to answer; it was about the Krylov–Safonov theory. Meditating about it in the days that followed, I noticed an immense amount of things I simply ignored – but had the impression that I should not. These notes arise from an effort to learn those things and master the material, as much as possible.

I thank Professor Eduardo Teixeira for having introduced me to the topic of fully nonlinear elliptic PDEs, as well as for a vastitude of interesting conversations. I express my heartfelt gratitude towards Professors Héctor Chang-Lara, Andrzej Święch and José Miguel Urbano, for wonderful discussions – on the topics in this book and beyond. Collaborating with them has been a genuinely challenging and stimulating experience, and a pleasant way to move forward in the profession; and I cannot be grateful enough.

I also thank Professors Mark Allen, Damião Araújo, Julián Bonder, Anne Bronzi, Disson dos Prazeres, Raimundo Leitão, Diego Moreira, Gabrielle

Nornberg, Júlio Rossi, Nicolau Saldanha, Henrik Shahgholian, Boyan Sirakov, and Mariana Smit Vega Garcia for their generosity to talk about mathematics with me.

A first idea on how to structure this set of notes crossed my mind when preparing to give a mini-course at the XII Americas Conference on Differential Equations and Nonlinear Analysis, in Guanajuato; I thank CIMAT for the kind hospitality, and Héctor Chang-Lara and Renato Iturriaga for the support during my visit.

I am also grateful to the graduate students and post doctoral interns Pêdra Andrade, Vincenzo Bianca, Ricardo Castillo, Julio Cesar Correa, Edison Cuba, Gerardo Huaroto, David Jesus, Giane Rampasso, Makson Santos, and Miguel Walker.

1

Elliptic Partial Differential Equations

This chapter is an introduction to elliptic partial differential equations (PDEs). We start with elementary notions and basic results. Then we proceed with a discussion on foundational regularity results for this class of equations.

1.1 Basic Definitions and Facts

We are concerned with second-order equations of elliptic type, whose general formulation is

$$G\left(x, u, Du, D^2u\right) = 0.$$

Here, $G = G(x, r, p, M)$ stands for a rule relating the Hessian of the unknown D^2u, its gradient Du, the unknown u itself and the spatial variable x. In the following, we examine examples of elliptic equations.

Example 1.1 (Linear equations in nondivergence form) Let $\Omega \subset \mathbb{R}^d$ be an open and connected subset of $\mathbb{R}^d$. Denote by $S(d)$ the space of $d \times d$ symmetric matrices. A linear equation in nondivergence form can be written as

$$- \operatorname{Tr}\left(A(x)D^2u\right) + b(x) \cdot Du + c(x)u = 0 \quad \text{in} \quad \Omega, \qquad (1.1)$$

where $A\colon \Omega \to S(d)$ is a matrix-valued map, $b\colon \Omega \to \mathbb{R}^d$ is a vector field, and $c\colon \Omega \to \mathbb{R}$ is a scalar function. In this concrete case, G takes the form

$$G(x, r, p, M) = -\operatorname{Tr}\left(A(x)M\right) + b(x) \cdot p + c(x)r.$$

If there are constants $0 < \lambda_0 \le \lambda \le \Lambda$ such that $A(x)$ satisfies

$$\lambda|\xi|^2 \le A(x)\xi \cdot \xi \le \Lambda|\xi|^2,$$

for every $\xi \in \mathbb{R}^d$, we say that $A(\cdot)$ is a (λ, Λ)-elliptic matrix. In this case, (1.1) is a uniformly elliptic equation. If instead it holds $0 \leq \lambda$, we say that (1.1) is merely elliptic. To make the distinction more apparent, suppose $d = 2$, make $b \equiv 0$ and $c \equiv 0$, and let $A(x)$ be given by

$$A(x) := \begin{bmatrix} x_1 & 0 \\ 0 & x_2 \end{bmatrix}.$$

The equation becomes

$$x_1 \frac{\partial^2 u}{\partial x_1 \partial x_1} + x_2 \frac{\partial^2 u}{\partial x_2 \partial x_2} = 0;$$

if we prescribe this problem in the first quadrant $\mathbb{R}_+^2$ it is elliptic, although not uniformly elliptic. However, if the equation is supposed to hold in a subset $\Omega \subset \mathbb{R}_+^2$ strictly away from the axis (i.e., with distance strictly greater than some $\lambda > 0$), the problem becomes uniformly elliptic.

In either case, the operator in (1.1) satisfies a monotonicity property with respect to the matrix M. In fact, fix $p \in \mathbb{R}^d$ and $r \in \mathbb{R}$ and let $M, N \in S(d)$ be such that $N - M \geq 0$. That is,

$$(N - M)\xi \cdot \xi \geq 0,$$

for every $\xi \in \mathbb{R}^d$. Then

$$G(x,r,p,M) - G(x,r,p,N) = \text{Tr}\,(A(x)(N - M)) \geq 0. \qquad (1.2)$$

If we set $b \equiv 0$ and $c \equiv 0$ in (1.1), we recover a purely second-order equation; if we further restrict the operator by requiring $A(x)$ to be the identity matrix I we get the Laplace operator.

Example 1.2 (Linear equations in divergence form)   A linear equation in the divergence form can be written as

$$-\frac{\partial}{\partial x_j}\left(a^{ij}(x)\frac{\partial u}{\partial x_i}\right) + \frac{\partial}{\partial x_j}(b^j(x)u) + c(x)u = 0 \quad \text{in} \quad \Omega, \qquad (1.3)$$

where $A(x) := (a^{ij}(x))_{i,j=1,...,d}$ is a matrix-valued map, $b(x) = (b^1(x), \ldots, b^d(x))$ is a vector field, and $c(x)$ is a scalar function; $A(\cdot)$, $b(\cdot)$, and $c(\cdot)$ are defined over the domain Ω. An alternative way to write the operator in (1.3) is

$$-\text{div}(A(x)Du) + \text{div}(b(x)u) + c(x)u = 0.$$

As before, the ellipticity of the problem is governed by the behavior of the matrix A. The equation is elliptic if A is nonnegative, whereas it is uniform elliptic if

$$\lambda|\xi|^2 \le A(x)\xi \cdot \xi \le \Lambda|\xi|^2,$$

for every $\xi \in \mathbb{R}^d$. If the coefficients $A(\cdot)$, b$(\cdot)$, and $c(\cdot)$ are regular enough – say, of class C^2 – it is possible to write (1.3) in the form of (1.1). In fact, in the case where we are allowed to compute the derivatives in (1.3), it becomes

$$- \operatorname{Tr} \left(A(x)D^2 u \right) + \tilde{b}(x) \cdot Du + \tilde{c}(x)u = 0 \quad \text{in} \quad \Omega,$$

where

$$\tilde{b}^i := b^i + \left(a_{ij} \right)_{x_j} \quad \text{and} \quad \tilde{c}(x) := c(x) + b^j_{x_j}.$$

By setting $A(x) \equiv I$, b $\equiv 0$, and $c \equiv 0$, once again we recover the Laplace operator

$$- \operatorname{div}(Du) = -\Delta u.$$

The analysis of Examples 1.1 and 1.2 suggests, we can perturb the Laplace operator in (at least) two ways. Set b $\equiv 0$ and $c \equiv 0$. Let $\varphi \in C^\infty_c(\Omega, \mathbb{R}^d)$ and consider

$$A_\varepsilon(x) := \begin{bmatrix} 1 + \varepsilon\varphi_1(x) & 0 & \cdots & 0 \\ 0 & 1 + \varepsilon\varphi_2(x) & \cdots & 0 \\ \vdots & \vdots & \ddots & \vdots \\ 0 & 0 & \cdots & 1 + \varepsilon\varphi_d(x) \end{bmatrix},$$

for $0 < \varepsilon \ll 1$, sufficiently small. For every $\varepsilon > 0$ one can detach from the Laplace operator either in a nondivergent form, by taking

$$L[u] := - \operatorname{Tr} \left(A_\varepsilon(x)D^2 u \right),$$

or in a divergent form, by considering

$$L[u] := - \operatorname{div}(A_\varepsilon(x)Du).$$

A question arising in that exercise concerns qualitative properties one can transmit from the Laplace operator to the perturbed models governed by A_ε. A more subtle issue concerns the amount of information one can import, depending on the form of the perturbation (nondivergence or divergence form). We return to those questions further in the book.

In addition to the notion of classical solution – a function $u \in C^2(\Omega)$ satisfying an equation in the classical sense – we consider weak solutions to the PDEs under analysis. In the context of operators in the divergence form, we use the notion of weak solutions in the distributional sense.

Definition 1.3 (Weak distributional solutions)   Let $A \in L^2_{\text{loc}}(\Omega, S(d))$, $b \in L^2_{\text{loc}}(\Omega, \mathbb{R}^d)$, $c \in L^2_{\text{loc}}(\Omega)$, and $f \in L^1_{\text{loc}}(\Omega)$. We say that $u \in W^{1,2}_{\text{loc}}(\Omega)$ is a weak solution to

$$- \operatorname{div}(A(x)Du) + \operatorname{div}(b(x)u) + c(x)u = f \quad \text{in} \quad \Omega, \tag{1.4}$$

if, for every $\varphi \in C^\infty_c(\Omega)$, we have

$$\int_\Omega (A(x)Du - b(x)u) \cdot D\varphi + \int_\Omega c(x)u\varphi = \int_\Omega f\varphi. \tag{1.5}$$

We refer to (1.5) as the *weak form* of (1.4). The test function φ used in Definition 1.3 can be taken in different functional spaces when appropriate. To write the weak form (1.5) we only require the coefficients of the equation to be in appropriate L^2-spaces.

In the context of divergence form problems, it is sometimes convenient to write equations as

$$- \operatorname{div} a(x, u, Du) = 0 \quad \text{in} \quad \Omega, \tag{1.6}$$

where $a \colon \Omega \times \mathbb{R} \times \mathbb{R}^d \to \mathbb{R}^d$ is a suitable vector field. Here, the notion of weak distributional solution adjusts in the obvious way. For instance, suppose

$$|a(x, r, p)| \le C\big(1 + |r|^q + |p|^r\big),$$

for some $C > 0$ and $1 < q \le r$; hence, $u \in W^{1,r}_{\text{loc}}(\Omega)$ is a weak solution to (1.6) if

$$\int_\Omega a(x, u, Du) \cdot D\varphi = 0,$$

for every $\varphi \in C^\infty_c(\Omega)$. The formulation in (1.6) allows us to consider diffusion coefficients depending also on u and Du, even in a nonlinear way. This observation leads us to our next example.

Example 1.4 (The p-Laplace operator)   Let $p > 1$ be fixed and consider

$$\operatorname{div}\big(|Du|^{p-2}Du\big) = f \quad \text{in} \quad \Omega, \tag{1.7}$$

for some function $f \in L^q(\Omega)$, with $q \in (1, \infty]$. We denote the operator in (1.7) by Δ_p and write the associated equation as $\Delta_p u = f$. The p-Laplace describes diffusion processes whose coefficients depend on the gradient of the solutions. At the points where Du vanishes, the ellipticity of the equation collapses. If $p > 2$, the problem degenerates, whereas in the case $p < 2$ it becomes singular.

Clearly, the case $p = 2$ recovers the Laplace operator. We observe that (1.7) is the Euler–Lagrange equation associated with the L^p-energy functional

$$I_p[u] := \int_\Omega \frac{|Du|^p}{p} + uf \, dx.$$

In analogy with the case $p = 2$, we sometimes refer to the solutions to $\Delta_p u = 0$ as p-harmonic functions.

Setting $p = 1$ in (1.7) we discover an intrinsically geometric equation. In fact, we get

$$\text{div} \left(\frac{1}{|Du|} Du \right) = 0$$

in Ω. For every $\alpha \in \mathbb{R}$, suppose the level set $\{u = \alpha\}$ is a smooth hypersurface; hence, its mean curvature is given by the divergence of the normal unit vector. The conclusion is that the level sets of u have zero mean curvature; i.e., $\Gamma_\alpha(u) := \{u = \alpha\}$ is a minimal surface for every $\alpha \in \mathbb{R}$.

Another instance of interest is the case $p \to \infty$; we examine it here from a heuristic viewpoint. Suppose $u \in C^2(\Omega)$ is p-harmonic and compute

$$0 = \text{div} \left(|Du|^{p-2} Du \right) = |Du|^{p-2} \Delta u + (p-2)|Du|^{p-4} u_{x_i} u_{x_j} u_{x_i x_j};$$

dividing both sides of the former equality by $|Du|^{p-2}(p-2)$ we obtain

$$\frac{\Delta u}{p-2} + \frac{1}{|Du|^2} u_{x_i} u_{x_j} u_{x_i x_j} = 0.$$

As $p \to \infty$ the operator becomes

$$\Delta_\infty u = D^2 u Du \cdot Du = 0.$$

Similar to what happens in the case $1 < p < \infty$, the ∞-Laplacian can also be regarded as an Euler–Lagrange equation. As one could expect, it relates to an L^∞-type of energy. Indeed, let $w \colon \partial\Omega \to \mathbb{R}$ be a Lipschitz continuous function. We seek $u \colon \Omega \to \mathbb{R}$ coinciding with w on $\partial\Omega$ and such that

$$\underset{x \in \Omega'}{\text{ess sup}} |Du(x)| \leq \underset{x \in \Omega'}{\text{ess sup}} |Dw(x)|,$$

for every open set $\Omega' \subset \Omega$.

We note the notion of weak distributional solutions is appropriate for the p-Laplace operator. However, it does not seem useful when considering the ∞-Laplace. In general, when turning our attention to nondivergent equations, as in Example 1.1, the notion of a weak distributional solution is not adequate. In this context we resort to the concept of the viscosity solution.

For ease of presentation, consider an operator $F \colon \Omega \times \mathbb{R} \times \mathbb{R}^d \times S(d) \to \mathbb{R}$. Suppose that for (x, r, p) arbitrary and $M, N \in S(d)$, with $M - N \geq 0$, we have

$$F(x, p, r, M) \leq F(x, p, r, N). \tag{1.8}$$

An operator satisfying (1.8) is called *degenerate elliptic*.

Definition 1.5 (*C*-viscosity solutions)   Suppose $F \in C(\Omega \times \mathbb{R} \times \mathbb{R}^d \times S(d))$. A function $u \in \mathrm{USC}(\Omega)$ is a *C*-viscosity subsolution to

$$F\big(x, u, Du, D^2 u\big) = 0 \quad \text{in} \quad \Omega, \tag{1.9}$$

if, for every $\varphi \in C^2(\Omega)$ and $x_0 \in \Omega$ such that $(u - \varphi)(x_0) \geq (u - \varphi)(x)$ locally, we have

$$F\big(x_0, u(x_0), D\varphi(x_0), D^2 \varphi(x_0)\big) \leq 0.$$

Conversely, we say that $u \in \mathrm{LSC}(\Omega)$ is a *C*-viscosity supersolution to (1.9) if, for every $\varphi \in C^2(\Omega)$ and $x_0 \in \Omega$ such that $(u - \varphi)(x_0) \leq (u - \varphi)(x)$ locally, we have

$$F\big(x_0, u(x_0), D\varphi(x_0), D^2 \varphi(x_0)\big) \geq 0.$$

Finally, if $u \in C(\Omega)$ is both a *C*-viscosity subsolution and a supersolution to (1.9), we say it is a *C*-viscosity solution to (1.9).

The notion of viscosity solutions recasts important properties of classical solutions in the context where C^2-regularity is not available. For example, suppose $F = F(M)$ is a degenerate elliptic operator and $v \in C^2(\Omega)$ is a classical solution to

$$F\big(D^2 v\big) = 0 \quad \text{in} \quad \Omega.$$

Take $\varphi \in C^2(\Omega)$ and $x_0 \in \Omega$ such that $v - \varphi$ attains a maximum at x_0. Clearly $D^2 v(x_0) - D^2 \varphi(x_0) \leq 0$. Hence,

$$F\big(D^2 \varphi(x_0)\big) \leq F\big(D^2 v(x_0)\big) = 0,$$

and the test function φ satisfies the inequality characterizing *C*-viscosity subsolutions. We will discuss further properties of *C*-viscosity solutions throughout the book. We proceed with an example.

Example 1.6 (Pucci extremal operators)   Let $0 < \lambda \leq \Lambda$ be fixed constants. Define the operators $\mathcal{P}^{\pm}_{\lambda, \Lambda} \colon S(d) \to \mathbb{R}$ as

$$\mathcal{P}^{+}_{\lambda, \Lambda}(M) := -\lambda \operatorname{Tr}\big(M^+\big) + \Lambda \operatorname{Tr}\big(M^-\big)$$

and

$$\mathcal{P}^-_{\lambda,\Lambda}(M) := -\Lambda \operatorname{Tr}\left(M^+\right) + \lambda \operatorname{Tr}\left(M^-\right),$$

where M^+ is the positive semidefinite part of the matrix M and M^- is its negative semidefinite part. Although $\mathcal{P}^\pm_{\lambda,\Lambda}$ are nonlinear, a simple calculation shows these operators are positively homogeneous of degree one. As a consequence, we learn that $\mathcal{P}^\pm_{\lambda,\Lambda}$ can not be differentiable operators. Indeed, were $\mathcal{P}^\pm_{\lambda,\Lambda}$ differentiable, their gradients would be homogeneous of degree zero and, therefore, constant maps. From this fact, we would infer the extremal operators are linear. By setting $\lambda = \Lambda = 1$, we have $\mathcal{P}^\pm_{\lambda,\Lambda}(M) = \operatorname{Tr} M$ and recover the Laplace operator.

Pucci operators play a central role in the analysis of nonlinear elliptic equations. They are used to characterize classes of viscosity solutions and in the statement of results aiming at holding for general, abstract operators. We make extensive use of extremal operators in the next chapters. We continue with yet another example.

Example 1.7 (Fully nonlinear elliptic operator)   Let $\mathcal{N} \subset \Omega$ be a null set, with respect to the Lebesgue measure. Let $F : (\Omega \setminus \mathcal{N}) \times \mathbb{R} \times \mathbb{R}^d \times S(d) \to \mathbb{R}$ be a measurable function. For constants $0 < \lambda \le \Lambda$, suppose the operator $F = F(x,r,p,M)$ satisfies

$$F(x,r,p,M) - \Lambda \operatorname{Tr}(P) \le F(x,r,p,M+P) \le F(x,r,p,M) - \lambda \operatorname{Tr}(P), \tag{1.10}$$

for every $(x,r,p) \in (\Omega \setminus \mathcal{N}) \times \mathbb{R} \times \mathbb{R}^d$ and $M, P \in S(d)$, with $P \ge 0$. Roughly speaking, (1.10) says that moving from a symmetric matrix M in the positive direction P, the operator changes proportionally to P itself. Such changes are controlled by the constants λ and Λ. Condition (1.10) is called uniform ellipticity. An operator F satisfying both inequalities in (1.10) is said to be (λ, Λ)-elliptic, or uniformly elliptic.

The class of operators introduced in Example 1.7 includes a number of important examples. For instance, the model in Example 1.1 satisfies (1.10), as

$$-\operatorname{Tr}(A(x)(M+P)) + \operatorname{Tr}(A(x)M) = -\operatorname{Tr}(A(x)P) \ge -\Lambda \operatorname{Tr}(P),$$

provided A is a (λ, Λ)-elliptic matrix and $M, P \in S(d)$, with $P \ge 0$. The remainder inequality follows similarly. The Pucci extremal operators also satisfy (1.10); notice that for any two symmetric matrices M and P, we have $M + P = M^+ + P - M^-$. Hence

$$(M+P)^+ = M^+ + P^+ \quad \text{and} \quad (M+P)^- = M^-.$$

As a result,

$$\begin{aligned}
\mathcal{P}^+_{\lambda,\Lambda}(M+P) &= -\lambda\,\mathrm{Tr}\left[(M+P)^+\right] + \Lambda\,\mathrm{Tr}\left[(M+P)^-\right] \\
&= -\lambda\,\mathrm{Tr}\left(M^+ + P^+\right) + \Lambda\,\mathrm{Tr}\left(M^-\right) \\
&= -\lambda\,\mathrm{Tr}\left(M^+\right) + \Lambda\,\mathrm{Tr}\left(M^-\right) - \lambda\,\mathrm{Tr}\left(P^+\right) \\
&= \mathcal{P}^+_{\lambda,\Lambda}(M) - \lambda\,\mathrm{Tr}(P).
\end{aligned}$$

Also

$$\begin{aligned}
\mathcal{P}^+_{\lambda,\Lambda}(M) - \Lambda\,\mathrm{Tr}(P) &= -\lambda\,\mathrm{Tr}\left(M^+ + P^+ - P^+\right) + \Lambda\,\mathrm{Tr}\left(M^- + P^-\right) - \Lambda\,\mathrm{Tr}\left(P^+\right) \\
&= \mathcal{P}^+_{\lambda,\Lambda}(M+P) + (\lambda - \Lambda)\,\mathrm{Tr}(P) \\
&\leq \mathcal{P}^+_{\lambda,\Lambda}(M+P).
\end{aligned}$$

We notice that (1.10) implies degenerate ellipticity. In fact, if $F = F(x,r,p,M)$ satisfies (1.10) and $M - N \geq 0$, it follows that

$$\begin{aligned}
F(x,r,p,M) = F(x,r,p,N+M-N) &\leq F(x,r,p,N) \\
&- \lambda\,\mathrm{Tr}(M-N) \leq F(x,r,p,N).
\end{aligned}$$

An equivalent characterization of uniform ellipticity is possible in terms of matrices' norms. We say the operator $F: S(d) \to \mathbb{R}$ is (λ,Λ)-elliptic if

$$\lambda\|N\| \leq F(M) - F(M+N) \leq \Lambda\|N\|,$$

for every $M, N \in S(d)$, with $N \geq 0$. The following inequality will be useful further.

Lemma 1.8 *Let $M, N \in S(d)$ and suppose $F: S(d) \to \mathbb{R}$ is a (λ,Λ)-elliptic operator. Then,*

$$F(M) \leq F(M+N) + \Lambda\|N^+\| - \lambda\|N^-\|,$$

for every $M, N \in S(d)$.

Proof For every $M, N = N^+ - N^-$ we have

$$F(M+N) \geq F\left(M+N^+\right) + \lambda\|N^-\|.$$

Moreover,

$$F\left(M+N^+\right) \geq F(M) - \Lambda\|N^+\|.$$

By combining both inequalities the result follows.   □

As a consequence of Lemma 1.8 we derive an upper bound for the difference of any two matrices in $F^{-1}(\{0\})$.

Corollary 1.9 *Let $M, N \in S(d)$ be such that $F(M) = F(N) = 0$. Then*

$$\|M - N\| \leq \frac{\Lambda + \lambda}{\lambda} \|(M - N)^+\| = \frac{\Lambda + \lambda}{\lambda} \sup_{\theta \in \mathbb{S}^{d-1}} \left(\theta^T (M - N)\theta\right)^+.$$

Proof It follows from Lemma 1.8 that

$$0 = F(M) = F(M - N + N)$$
$$\leq F(N) + \Lambda \|(M - N)^+\| - \lambda \|(M - N)^-\|;$$

i.e.,

$$\lambda \|(M - N)^-\| \leq \Lambda \|(M - N)^+\|.$$

By adding $\lambda \|(M - N)^+\|$ on both sides of the former inequality and applying the triangle inequality we get

$$\lambda \|M - N\| \leq (\Lambda + \lambda) \|(M - N)^+\| = (\Lambda + \lambda) \sup_{\theta \in \mathbb{S}^{d-1}} \left(\theta^T (M - N)\theta\right)^+.$$

$\square$

An additional monotonicity condition involves the dependence of F on the zeroth-order term. From now on, we denote by $\mathcal{N}$ the null subset of Ω comprising the points where $F(\cdot, r, p, M)$ is not defined.

Definition 1.10 (Properness) Let $F \colon (\Omega \setminus \mathcal{N}) \times \mathbb{R} \times \mathbb{R}^d \times S(d) \to \mathbb{R}$ be a measurable function. We say that $F = F(x, r, p, M)$ is proper if it is degenerate elliptic and for every $(x, p, M) \in (\Omega \setminus \mathcal{N}) \times \mathbb{R}^d \times S(d)$ we have

$$F(x, r, p, M) \leq F(x, s, p, M),$$

whenever $r \leq s$.

The operator in Example 1.7 is required to be merely measurable, with no further assumptions (e.g., on its continuity). More subtle is the fact that $F(\cdot, r, p, M)$ is defined over a set of full measure; in particular, there might exist points $x_0 \in \Omega$ where the operator is not well-defined. This scenario challenges the notion of C-viscosity solutions, as it might not be possible to test the inequalities involved in the definition at some particular point $x_0 \in \Omega$ – simply because the operator might not be defined at that point. To circumvent these difficulties – and also to address further issues in the realm of viscosity solutions – we resort to the notion of the L^p-viscosity solution, introduced by Caffarelli et al. (1996).

The definition of the L^p-viscosity solution must be consistent with the one for C-viscosity solutions, in the sense that if the operator happens to be continuous, they should coincide. On the other hand, the former has to account for the fact that, a priori, the operator is defined almost everywhere in the domain Ω. The precise definition reads as follows.

Definition 1.11 (L^p-viscosity solution)   Let $F \colon (\Omega \setminus \mathcal{N}) \times \mathbb{R} \times \mathbb{R}^d \times S(d) \to \mathbb{R}$ be a measurable function. Suppose F is proper and let $f \in L^p(\Omega)$, for $p > d/2$. A function $u \in \mathrm{USC}(\Omega)$ is an L^p-viscosity subsolution to

$$F\left(x, u, Du, D^2 u\right) = f \quad \text{in} \quad \Omega, \tag{1.11}$$

if, for every $\varphi \in W_{\mathrm{loc}}^{2,p}(\Omega)$ such that there is $\varepsilon > 0$, and an open set $U \subset \Omega$ for which

$$F\left(x, \varphi(x), D\varphi(x), D^2\varphi(x)\right) - f(x) \geq \varepsilon$$

a.e.-$x \in U$, then $u - \varphi$ does not attain a local maximum in U. A function $u \in \mathrm{LSC}(\Omega)$ is an L^p-viscosity supersolution to (1.11) if for every $\varphi \in W_{\mathrm{loc}}^{2,p}(\Omega)$ such that there is $\varepsilon > 0$, and an open set $U \subset \Omega$ for which

$$F\left(x, \varphi(x), D\varphi(x), D^2\varphi(x)\right) - f(x) \leq -\varepsilon$$

a.e.-$x \in U$, then $u - \varphi$ does not attain a local minimum in U. If $u \in C(\Omega)$ is simultaneously an L^p-viscosity subsolution and a supersolution to (1.11), we say u is an L^p-viscosity solution to (1.11).

When dealing with the notion of L^p-viscosity solutions we require the operator F to satisfy a structure condition. To make matters precise, we state it in the form of a definition.

Definition 1.12 (Structure condition)   Let $F \colon (\Omega \setminus \mathcal{N}) \times \mathbb{R} \times \mathbb{R}^d \times S(d) \to \mathbb{R}$ be a measurable function such that $F(x, \cdot, \cdot, \cdot) \in L^p(\Omega)$, for some $p > d/2$. If there are constants $0 < \lambda \leq \Lambda$ and $\gamma > 0$, and a modulus of continuity $\omega_F \colon \mathbb{R}_+ \to \mathbb{R}_+$ such that

$$\mathcal{P}_{\lambda,\Lambda}^-(M - N) - \gamma|p - q| - \omega_F\left((s - r)^+\right) \leq F(x, r, p, M) - F(x, s, q, N)$$
$$\leq \mathcal{P}_{\lambda,\Lambda}^+(M - N) + \gamma|p - q| - \omega_F\left((r - s)^+\right),$$

we say F satisfies a $(\lambda, \Lambda, \gamma, \omega_F)$-structure condition.

We notice the structure condition in Definition 1.12 enforces the uniform ellipticity of F. Indeed, take $M, P \in S(d)$ with $P \geq 0$. It follows from the structure condition that

$$F(x, r, p, M + P) - F(x, r, p, M) \leq \mathcal{P}_{\lambda,\Lambda}^+(P) = -\lambda \operatorname{Tr}(P+),$$

for every $(x, r, p) \in (\Omega \setminus \mathcal{N}) \times \mathbb{R} \times \mathbb{R}^d$. The remainder inequality follows similarly. Moreover, the structure condition implies properness; since degenerate ellipticity follows from the former observation, it remains for us to check that, $F = F(x, r, p, M)$ is nondecreasing with respect to r. For $r \leq s$, $\omega_F\big((r - s)^+\big) \geq 0$; hence,

$$F(x, r, p, M) \leq F(x, s, p, M) + \omega_F\big((r - s)^+\big),$$

and the conclusion follows. Finally, we notice the constant $\gamma > 0$ can be replaced with a function $\gamma \in L^\infty(\Omega)$.

A typical formulation falling within the scope of Definition 1.12 is

$$F(x, M) + H(x, p),$$

where F satisfies the structure condition and

$$H(x, p) \sim \gamma(x)|p|,$$

for some $\gamma \in L^\infty(\Omega)$.

Suppose $F = F(x, p, M)$ is a fully nonlinear operator satisfying the structure condition. Let $f \in L^p(\Omega)$, for some $p > d/2$ fixed, though arbitrary. If $u \in C(\Omega)$ is an L^p-viscosity solution

$$F\big(x, Du, D^2u\big) = f \quad \text{in} \quad \Omega, \tag{1.12}$$

it is an L^p-viscosity subsolution to

$$\mathcal{P}^-_{\lambda, \Lambda}\big(D^2u\big) - \gamma|Du| = f + F(x, 0, 0) \quad \text{in} \quad \Omega.$$

To verify this claim, it suffices to show that for every $\varphi \in W^{2,p}_{\text{loc}}(\Omega)$, if there exists $\varepsilon > 0$ and an open subset $U \subset \Omega$ such that

$$\mathcal{P}^-_{\lambda, \Lambda}\big(D^2\varphi(x)\big) - \gamma|D\varphi(x)| - f(x) - F(x, 0, 0) \geq \varepsilon \quad \text{almost everywhere in } \Omega,$$

then $u - \varphi$ cannot have a maximum point inside U. But it follows easily from the definition of L^p-viscosity solution and the structure condition, since

$$\mathcal{P}^-_{\lambda, \Lambda}\big(D^2\varphi(x)\big) - \gamma|D\varphi(x)| - f - F(x, 0, 0) \leq F\big(x, D\varphi(x), D^2\varphi(x)\big)$$
$$- f(x) - F(x, 0, 0).$$

An entirely analogous argument implies that an L^p-viscosity solution to (1.12) is an L^p-viscosity supersolution to

$$\mathcal{P}^+_{\lambda, \Lambda}\big(D^2u\big) + \gamma|Du| = f + F(x, 0, 0) \quad \text{in} \quad \Omega.$$

From the former implications, one can derive from the extremal operators information about a (very) large class of problems. As a matter of fact, let $u \in C(\Omega)$ be an L^p-viscosity solution to an arbitrary equation

$$G(x, Du, D^2u) = f \quad \text{in} \quad \Omega.$$

If G satisfies the structure condition, we know that u is a subsolution to

$$\mathcal{P}^-_{\lambda,\Lambda}(D^2u) - \gamma|Du| = f \quad \text{in} \quad \Omega \tag{1.13}$$

and a supersolution to

$$\mathcal{P}^+_{\lambda,\Lambda}(D^2u) + \gamma|Du| = f \quad \text{in} \quad \Omega. \tag{1.14}$$

As a consequence, by establishing results for subsolutions and supersolutions to (1.13) and (1.14) respectively, one recovers information available for *every* L^p-viscosity solution to *any* equation driven by an operator satisfying the structure condition. This discussion motivates a definition.

Definition 1.13 (Class of L^p-viscosity solutions)   Let $0 < \lambda \le \Lambda$ and $\gamma > 0$ be constants. We denote with $\underline{S}(\lambda, \Lambda, \gamma, f) \subset C(\Omega)$ the set of every L^p-viscosity subsolution to (1.13). Conversely, we denote with $\overline{S}(\lambda, \Lambda, \gamma, f) \subset C(\Omega)$ the set of every L^p-viscosity supersolution to (1.14). Finally, define

$$S(\lambda, \Lambda, \gamma, f) := \underline{S}(\lambda, \Lambda, \gamma, f) \cap \overline{S}(\lambda, \Lambda, \gamma, f).$$

Heuristically, $S(\lambda, \Lambda, \gamma, f)$ gathers every L^p-viscosity solution to any equation governed by operators satisfying the structure condition. Usually we write $S(\lambda, \Lambda, f) := S(\lambda, \Lambda, 0, f)$.

At this point we discuss a question that seems natural in the context of the classes $\underline{S}$ and $\overline{S}$, but whose answer will be available only further in Section 1.7.2. We have learned that if $u \in C(\Omega)$ solves an equation of the form $F(D^2u) = f$ in Ω, then it solves $\mathcal{P}^-_{\lambda,\Lambda}(D^2u) \le f$ and $\mathcal{P}^+_{\lambda,\Lambda}(D^2u) \ge f$ in Ω. The question concerns a form of reverse implication.

Suppose, conversely, that $w \in C(\Omega)$ solves

$$\mathcal{P}^-(D^2w) \le 0 \quad \text{and} \quad \mathcal{P}^+(D^2w) \ge 0.$$

We ask whether it is always possible to find an operator $\overline{F}: S(d) \to \mathbb{R}$, satisfying appropriate structure conditions, such that

$$\overline{F}(D^2w) = 0 \quad \text{in} \quad \Omega.$$

The answer to this question is negative, and it follows by combining results in regularity theory with counterexamples related to homogeneous equations of the form $F = 0$; we return to this problem later in Section 1.7.2.

When studying regularity results by approximation methods, a fundamental ingredient concerns the stability of the solutions. The question is posed as follows.

Let $(F_n)_{n \in \mathbb{N}}$ be a sequence of fully nonlinear operators $F_n \colon \Omega \setminus \mathcal{N} \times \mathbb{R} \times \mathbb{R}^d \times S(d) \to \mathbb{R}$; let $(u_n)_{n \in \mathbb{N}} \subset C(B_1)$ be a sequence of functions satisfying

$$F_n\big(x, u_n, Du_n, D^2 u_n\big) = 0 \quad \text{in} \quad \Omega$$

in the L^p-viscosity sense. Suppose F_n converges locally uniformly to some operator $F_\infty = F_\infty(x, r, p, M)$; suppose further $u_n \to u_\infty$ locally uniformly. The question is whether or not u_∞ is an L^p-viscosity solution to

$$F_\infty\big(x, u_\infty, Du_\infty, D^2 u_\infty\big) = 0 \quad \text{in} \quad \Omega.$$

The answer to this question is indeed positive. To establish this fact we resort to a further notion of solution and to an existence result. This is the content of Section 1.1.1.

1.1.1 Stability of Viscosity Solutions

In what follows we prove a stability result for L^p-viscosity solutions. A crucial step in the poof is the existence of strong solutions to an equation driven by the extremal operators. We proceed with the following definition.

Definition 1.14 (Strong solutions) Let $F \colon \Omega \times \mathbb{R} \times \mathbb{R}^d \times S(d) \to \mathbb{R}$ and $p > d/2$. We say that $u \in W^{2,p}_{\mathrm{loc}}(\Omega)$ is an L^p-strong solution to

$$F\big(x, u, Du, D^2 u\big) = 0 \quad \text{in} \quad \Omega,$$

if $F\big(x, u(x), Du(x), D^2 u(x)\big) = 0$ for almost every $x \in \Omega$.

The requirement $p > d/2$ is to ensure $u \in C(\Omega)$. Throughout these notes we refer to some constants as *universal*. In general, a constant is said to be universal if it depends only on the dimension d, and the ellipticity (λ, Λ). In different contexts, we abuse terminology and also refer as *universal* constants depending on other quantities, provided they are intrinsic to the problem under analysis. In those cases, we make explicit mention.

We also emphasize the range in which the exponent p varies. Although in Definitions 1.11 and 1.14 we required $p > d/2$, this condition is insufficient to establish a soundly based theory; see, for instance, Koike and Święch (2012,

p. 2) for its importance in the context of local maximum principles. In what follows, we impose $p > p_0$, where $p_0 = p_0(d, \lambda, \Lambda)$ is Escauriaza's exponent introduced in Escauriaza (1993). Next we recall a definition used further in this section.

Definition 1.15 (Uniform exterior cone condition) Let $\Omega \subset \mathbb{R}^d$ be open and bounded. We say that Ω satisfies a uniform exterior cone condition if there exists $r, \theta > 0$ such that for every $x \in \partial\Omega$ one find a cone C of opening θ and vertex at the origin satisfying

$$(x + C) \cup B_r(x) \subset \mathbb{R}^d \setminus \Omega.$$

From an informal perspective, this condition can be phrased as follows: Given $\Omega \subset \mathbb{R}^d$, there exists a cone with fixed opening touching $\partial\Omega$ from outside the domain at every $x \in \partial\Omega$. We proceed with a result on the existence of L^p-strong solutions.

Proposition 1.16 (Existence of L^p-strong solutions) *Suppose $\Omega \subset \mathbb{R}^d$ satisfies a uniform exterior cone condition. Let $f \in L^p(\Omega)$ and $g \in C(\partial\Omega)$, for $p > p_0$. There exist $u, v \in W^{2,p}_{loc}(\Omega) \cap C(\overline{\Omega})$ satisfying*

$$\mathcal{P}^+\left(D^2 u\right) + \gamma |Du| \leq f \quad in \quad \Omega$$

and

$$\mathcal{P}^-\left(D^2 v\right) - \gamma |Dv| \geq f \quad in \quad \Omega$$

in the L^p-strong sense, with $u = v = g$ on $\partial\Omega$. In addition, there exists a constants $C_1 > 0$ such that the functions u and v satisfy

$$\|u\|_{L^\infty(\Omega)}, \|v\|_{L^\infty(\Omega)} \leq \|g\|_{L^\infty(\partial\Omega)} + C_1 (\text{diam}(\Omega))^{2-\frac{d}{p}} \|f\|_{L^p(\Omega)}, \quad (1.15)$$

and $C_1 = C_1(d, \lambda, \Lambda, \gamma, \text{diam}(\Omega))$. Also, for every $\Omega' \Subset \Omega$, there exists $C_2 > 0$ such that

$$\|u\|_{W^{2,p}(\Omega')}, \|v\|_{W^{2,p}(\Omega')} \leq C \left(\|g\|_{L^\infty(\partial\Omega)} + \|f\|_{L^p(\Omega)} \right), \quad (1.16)$$

with $C_2 = C_2(d, \lambda, \Lambda, \gamma, p, \text{diam}(\Omega), \text{dist}(\Omega', \partial\Omega))$.

Proof We detail the case of subsolutions, the remaining one being entirely analogous. For ease of presentation, we split the proof into four steps.

Step 1. We consider a sequence of approximating problems and start by verifying (1.15) at the level of approximating solutions. Let $(f_n)_{n \in \mathbb{N}} \subset C^\infty(\Omega)$ be such that $f_n \to f$ in $L^p(\Omega)$ and $|f_n(x) - f(x)| \to 0$ for almost every $x \in \Omega$. The existence of $u_n \in C^2(\Omega) \cap C(\partial\Omega)$ satisfying

$$\mathcal{P}^+\left(D^2 u_n\right) + \gamma |Du_n| = f_n \quad in \quad \Omega,$$

agreeing with g on $\partial\Omega$, follows from standard results in the literature; see, for instance Gilbarg and Trudinger (2001, Theorem 17.17). Moreover, because u_n is a classical solution, we have

$$\|u_n\|_{L^\infty(\Omega)} \le \|g\|_{L^\infty(\partial\Omega)} + C_1(\text{diam}(\Omega))^{2-\frac{d}{p}}\|f_n\|_{L^p(\Omega)}. \qquad (1.17)$$

Step 2. Now we examine (1.16) in the context of approximating solutions. We start with a covering argument. For every $\Omega' \Subset \Omega$ and $0 < r < 1$, there exists a finite family of open balls $B_r(x_1), B_r(x_2), \ldots, B_r(x_M)$, with $x_i \in \Omega'$ for every $i = 1, \ldots, M$, such that

$$\Omega' \subset \left(\bigcup_{i=1}^{M} B_{r/2}(x_i)\right)$$

and

$$\overline{B_r(x_i)} \subset \Omega \qquad \text{for} \quad i = 1, \ldots, M.$$

Therefore, it suffices to derive (1.16) in $B_{1/2}$.

Now we let $\rho \in (0,1)$. Choose $\eta \in C_0^2(B_r)$ satisfying $0 \le \eta \le 1$, and such that

$$\eta \equiv 1 \quad \text{in} \quad B_{\rho r} \qquad \text{and} \qquad \eta \equiv 0 \quad \text{in} \quad B_r \setminus B_{\overline{\rho}r},$$

for

$$\overline{\rho} := \frac{1+\rho}{2}.$$

Refine the choice of η to ensure

$$|D\eta| \le \frac{4}{(1-\rho)r} \qquad \text{and} \qquad \|D^2\eta\| \le \frac{16}{((1-\rho)r)^2}. \qquad (1.18)$$

Notice that $\eta u_n \in C_0^2(B_{\overline{\rho}r})$ and

$$D^2(\eta u_n) = \eta D^2 u_n + 2D\eta \otimes Du_n + u_n D^2\eta.$$

In addition, extend ηu_n outside of $B_{\overline{\rho}r}$ by zero and define $\overline{f}(x) := \mathcal{P}^+(D^2(\eta u_n)(x))$. It follows that ηu_n is a C-viscosity solution to

$$\mathcal{P}^+(D^2 w) = \overline{f} \quad \text{in} \quad B_r,$$

with $w = 0$ on ∂B_r. We learn from Caffarelli (1989) that

$$\|D^2(\eta u_n)\|_{L^p(B_{\overline{\rho}r})} \le C\|\overline{f}\|_{L^p(B_{\overline{\rho}r})},$$

for some universal constant $C > 0$; see also Caffarelli and Cabré (1995, Chapter 7). As a consequence, we recover

$$\begin{aligned}
\left\| D^2 u_n \right\|_{L^p(B_{\rho r})} \leq \left\| D^2(\eta u_n) \right\|_{L^p(B_{\bar{\rho} r})} &\leq C \left\| \mathcal{P}^+ \left(D^2(\eta u_n) \right) \right\|_{L^p(B_{\bar{\rho} r})} \\
&\leq C \left\| \mathcal{P}^+ \left(\eta D^2 u_n + 2 D\eta \otimes D u_n + u_n D^2 \eta \right) \right\|_{L^p(B_{\bar{\rho} r})} \\
&\leq C(\gamma) \left(\left\| f_n \right\|_{L^p(B_{\bar{\rho} r})} + \frac{\left\| D u_n \right\|_{L^p(B_{\bar{\rho} r})}}{(1-\rho)r} + \frac{\left\| u_n \right\|_{L^p(B_{\bar{\rho} r})}}{((1-\rho)r)^2} \right),
\end{aligned}$$
(1.19)

where $C(\gamma) = C(C, \gamma)$.

Step 3. In what follows we use an interpolation inequality to refine the estimate in (1.19). In particular, we are interested in removing the dependence on the L^p-norm of $D u_n$. For $k \in \mathbb{N}$ consider

$$\Psi_k(w) := \sup_{\rho \in (0,1)} (1-\rho)^k r^k \left\| D^k w \right\|_{L^p(B_{\rho r})}.$$

We can rewrite the last inequality in (1.19) in terms of Ψ_k; in fact,

$$\Psi_2(u_n) \leq C \left(r^2 \left\| f_n \right\|_{L^p(B_r)} + \Psi_1(u_n) + \Psi_0(u_n) \right).$$

Standard interpolation inequalities yield

$$\Psi_1(u_n) \leq \varepsilon \Psi_2(u_n) + \frac{C}{\varepsilon} \Psi_0(u_n),$$

for every $0 < \varepsilon \ll 1$. By choosing ε small enough, it follows that

$$\left\| D^2 u_n \right\|_{L^p(B_{\rho r})} \leq C \left(\left\| u_n \right\|_{L^\infty(B_r)} + \left\| f_n \right\|_{L^p(B_r)} \right).$$

We conclude that $(u_n)_{n \in \mathbb{N}}$ is uniformly bounded in $W^{2,p}_{\text{loc}}(B_r)$. As a consequence, there exists the weak limit u_∞ such that $u_n \to u_\infty$ in $W^{2,p}_{\text{loc}}(B_r)$. Firstly, the lower semicontinuity of the norm yields

$$\begin{aligned}
\left\| D^2 u_\infty \right\|_{L^p(B_{\rho r})} &\leq \liminf_{n \to \infty} \left\| D^2 u_n \right\|_{L^p(B_{\rho r})} \\
&\leq C \liminf_{n \to \infty} \left(\left\| u_n \right\|_{L^\infty(B_r)} + \left\| f_n \right\|_{L^p(B_r)} \right) \\
&\leq C \left(\left\| u_\infty \right\|_{L^\infty(B_r)} + \left\| f \right\|_{L^p(B_r)} \right).
\end{aligned}$$

By noticing that $\left\| u_\infty \right\|_{L^\infty(B_r)} \leq \left\| g \right\|_{L^\infty(\partial B_r)}$, the estimate in (1.16) follows. Revisiting (1.17) we conclude u_∞ also satisfies (1.15). Finally, the weak lower semicontinuity of the extremal operator $\mathcal{P}^+$ implies

$$\mathcal{P}^+ \left(D^2 u_\infty \right) + \gamma \left| D u_\infty \right| \leq \liminf_{n \to \infty} \left(\mathcal{P}^+ \left(D^2 u_n \right) + \gamma |D u_n| \right) \leq f.$$

It remains for us to verify that $u_\infty \in C(\overline{B_r})$.

Step 4. We prove that $u_\infty \in C(\overline{B_r})$ by verifying that $(u_n)_{n\in\mathbb{N}}$ is a Cauchy sequence in that space. Because $\mathcal{P}^+$ is subadditive, we get

$$f_n = \mathcal{P}^+\big(D^2(u_n - u_m + u_m)\big) + \gamma\big|D(u_n - u_m + u_m)\big|$$
$$\leq \mathcal{P}^+\big(D^2(u_n - u_m)\big) + \gamma\big|D(u_n - u_m)\big| + f_m,$$

which is

$$f_n - f_m \leq \mathcal{P}^+\big(D^2(u_n - u_m)\big) + \gamma\big|D(u_n - u_m)\big|.$$

Noticing that $u_n = u_m = g$ on ∂B_r and reversing the order of u_n and u_m, an application of standard results for classical solutions leads to

$$\sup_{B_r}(u_n - u_m)^-, \ \sup_{B_r}(u_m - u_n)^- \leq C\|f_n - f_m\|_{L^p(B_r)} \to 0$$

as $n, m \to \infty$. Hence $\|u_n - u_m\|_{L^\infty(B_r)} \to 0$ as $n, m \to \infty$ and the sequence is Cauchy in $C(\overline{B_r})$, which completes the proof. □

The next theorem concerns the stability of L^p-viscosity solutions.

Theorem 1.17 (Stability of viscosity solutions) *Let $p > p_0$. Let $(F_n)_{n\in\mathbb{N}}$, $(f_n)_{n\in\mathbb{N}} \subset L^p(\Omega)$ and $(u_n)_{n\in\mathbb{N}} \subset C(\Omega)$ be sequences such that:*

(i) *For every $n \in \mathbb{N}$, the operator $F_n\colon (\Omega \setminus \mathcal{N}) \times \mathbb{R} \times \mathbb{R}^d \times S(d) \to \mathbb{R}$ satisfies a $(\lambda, \Lambda, \gamma, \omega)$-structure condition.*
(ii) *For every $n \in \mathbb{N}$, the function u_n is an L^p-viscosity solution to*

$$F_n\big(x, u_n, Du_n, D^2 u_n\big) = f_n \quad in \quad \Omega.$$

Suppose there exists $u_\infty \in C(\Omega)$ such that $u_n \to u_\infty$ locally uniformly as $n \to \infty$. Suppose further there are F_∞ and f_∞ such that, for every $B_r(x_0) \subset \Omega$ and $\phi \in W^{2,p}(B_r(x_0))$ the function

$$g_n(x) := F_n\big(x, u_n(x), D\phi(x), D^2\phi(x)\big) - F_\infty\big(x, u_\infty(x), D\phi(x), D^2\phi(x)\big)$$
$$+ f_\infty(x) - f_n(x)$$

$$(1.20)$$

converges to zero in $L^p(B_r(x_0))$, as $n \to \infty$. Then u_∞ is an L^p-viscosity solution to

$$F_\infty\big(x, u_\infty, Du_\infty, D^2 u_\infty\big) = f_\infty \quad in \quad \Omega.$$

Proof We only prove the subsolution part, the supersolution case being entirely analogous. We proceed through a contradiction argument. First we suppose one can craft a particular sequence of functions; it leads to a contradiction. To conclude, we prove it is possible to produce the required sequence of functions.

Step 1. Seeking a contradiction, we start by supposing that u_∞ is not an L^p-viscosity subsolution to $F_\infty = f_\infty$. If this is the case, we can find $x^* \in \Omega$, numbers $\varepsilon^*, \delta^*, r > 0$, and a function $\varphi \in W^{2,p}(B_r(x^*))$ such that

$$F_\infty(x, u_\infty, D\varphi, D^2\varphi) > f + \varepsilon^* \quad \text{and} \quad B_r(x^*)$$

and

$$u_\infty(x^*) = \varphi(x^*), \tag{1.21}$$

but

$$u_\infty - \varphi < -\delta^* \quad \text{on} \quad \partial B_r(x^*). \tag{1.22}$$

At this point, we search for a sequence $(\varphi_n)_{n\in\mathbb{N}} \subset W^{2,p}(B_r(x^*)) \cap C(\overline{B_r(x^*)})$ satisfying

$$F_n(x, u_n(x), D(\varphi + \varphi_n), D^2(\varphi - \varphi_n)) \geq f_n(x) + \varepsilon^* \quad \text{in} \quad B_r(x^*). \tag{1.23}$$

If we produce such a sequence, we get a contradiction.

Step 2. Indeed, suppose we have produced such sequence $(\varphi_n)_{n\in\mathbb{N}}$. Consider $u_n - (\varphi + \varphi_n)$. Because $u_n \to u_\infty$ and $\varphi_n \to 0$ locally uniformly in $B_r(x^*)$, (1.21) and (1.22) ensure that $u_n - (\varphi + \varphi_n)$ has an interior maximum, for large values of $n \gg 1$. Because of (1.23), u_n can not be a viscosity subsolution to $F_n = f_n$, which is a contradiction. It remains for us to prove the existence of $(\varphi_n)_{n\in\mathbb{N}}$.

Step 3. We combine the structure condition satisfied by F_n and the fact that u_n is an L^p-viscosity subsolution to $F_n = f_n$ to get

$$F_n(x, u_n(x), D(\varphi + \varphi_n), D^2(\varphi + \varphi_n)) - f_n(x)$$
$$\geq F(x, u, D\varphi, D^2\varphi) - f(x) + \mathcal{P}^-(D^2\varphi_n) - \gamma|D\varphi_n| + g_n(x)$$
$$\geq \varepsilon^* + \mathcal{P}^-(D^2\varphi_n) - \gamma|D\varphi_n| + g_n(x).$$

Let φ_n be the strong solution to

$$\mathcal{P}^-(D^2w) - \gamma|Dw| \geq (-g_m)^+ \quad \text{in} \quad B_r(x^*),$$

with $\varphi_n = 0$ on $\partial B_r(x^*)$, whose existence is guaranteed by Proposition 1.16. For such choice of φ, (1.23) is ensured. Also, because $g_n \to 0$ in L^p, we have $\varphi_n \to 0$ in $L^\infty(B_r(x^*))$ and the proof is finished. $\square$

Remark 1.18 We emphasize that if $F_n \to F_\infty$ and $f_n \to f_\infty$ locally in $L^p(\Omega)$, the functions g_n defined in (1.20) satisfy the condition of Theorem 1.17.

In some applications, the sequence $(F_n)_{n \in \mathbb{N}}$ comprises (λ, Λ)-elliptic operators $F_n = F_n(M)$. In this case, F_n is Lipschitz continuous, with a universal modulus of continuity, depending only on d, λ, and Λ. Hence, there exists a (λ, Λ)-elliptic operator F_∞ such that $F_n \to F_\infty$ locally uniformly in $S(d)$, through a subsequence if necessary.

In this setting, we are interested in the subsequential limits of the family $(u_n)_{n \in \mathbb{N}}$; this family comprises the L^p-viscosity solutions to $F_n(D^2 u_n) = f_n$. If such subsequential limits exist in the (locally) uniform topology, choose one and denote it by u_∞. Under the additional condition that f_n converges to some f_∞ in an appropriate space, our goal is to conclude that $F_\infty(D^2 u_\infty) = f_\infty$ in the interior of the domain, provided appropriate boundary conditions are met.

We close this section with the Aleksandrov–Bakelman–Pucci maximum principle. The motivation to include it here is two-fold. First, it accounts for a result holding in the generality of the viscosity class $S(\lambda, \Lambda, \gamma, f)$ and illustrates the use of some concepts introduced before. Secondly, it is of paramount relevance in the study of regularity theory for problems in the nondivergence form.

1.1.2 The ABP Maximum Principle

In what follows we prove the Aleksandrov–Bakelman–Pucci (ABP) estimate. In brief, it controls the value of solutions in Ω in terms of their values on the boundary and a contribution of the source term f. One remarkable aspect of the estimate is that f contributes through its L^d-norm computed on the contact set with the appropriate envelope (convex in the case of supersolutions and concave in the case of subsolutions). We continue with a definition.

Definition 1.19 (Concave and convex envelopes) Let $u \in C(\Omega)$. The concave envelope of u in the domain Ω is the function $\Gamma_u^+ : \Omega \to \mathbb{R}$ defined as

$$\Gamma_u^+(x) := \inf \left\{ h(x) := \mathrm{a} + \mathrm{b} \cdot x \mid \mathrm{a} \in \mathbb{R}, \mathrm{b} \in \mathbb{R}^d \text{ and } u(x) \leq \mathrm{a} + \mathrm{b} \cdot x \right\}$$

Conversely, the convex envelope of u is the function $\Gamma_u^- : \Omega \to \mathbb{R}$ defined as

$$\Gamma_u^-(x) := \sup \left\{ h(x) := \mathrm{a} + \mathrm{b} \cdot x \mid \mathrm{a} \in \mathbb{R}, \mathrm{b} \in \mathbb{R}^d \text{ and } u(x) \geq \mathrm{a} + \mathrm{b} \cdot x \right\}$$

We are interested in the contact sets $\{u = \Gamma_u^{\pm}\}$, since the contribution of f to the ABP is given in terms of $\|f\|_{L^d(\{u=\Gamma_u^-\})}$ in the case of subsolutions, and $\|f\|_{L^d(\{u=\Gamma_u^+\})}$ in the case of supersolutions. To ease notation, we set

$$K^{\pm}(u, \Omega) := \{x \in \Omega \mid u(x) = \Gamma_u^{\pm}(x)\}.$$

A useful characterization for $K^{\pm}(u, U)$ is the following, where $U \subset \Omega$ is an arbitrary open set:

$$K^+(u, U) = \{x \in U \mid \exists p \in \mathbb{R}^d \text{ such that } u(y) \leq u(x) + p \cdot (y - x) \ \forall y \in U\}$$

and

$$K^-(u, U) = \{x \in U \mid \exists p \in \mathbb{R}^d \text{ such that } u(y) \geq u(x) + p \cdot (y - x) \ \forall y \in U\}.$$

Following closely Caffarelli et al. (1996), we use $K^{\pm}(u) := K^{\pm}(u, \Omega)$. It is useful to refer to K^+ as upper contact set and to K^- as lower contact set. Next we state the Aleksandrov–Bakelman–Pucci estimate.

Proposition 1.20 (Aleksandrov–Bakelman–Pucci)   *Let* $u \in C(\overline{\Omega})$ *be an* L^d*-viscosity solution to*

$$\mathcal{P}^-(D^2 u) - \gamma|Du| \leq f \quad in \quad \Omega \cap \{u > 0\}. \tag{1.24}$$

Then there exists $C > 0$ *such that*

$$\sup_{\Omega} u \leq \sup_{\partial\Omega} u^+ + C \operatorname{diam}(\Omega) \|f^+\|_{L^d(K^+(u))};$$

in addition, $C = C(d, \lambda, \Lambda, \operatorname{diam}(\Omega))$.

One interesting aspect of this estimate is that it controls a pointwise quantity by a measure-theoretic information. Next we prove Proposition 1.20; our presentation follows closely the one in Caffarelli et al. (1996); see also Gilbarg and Trudinger (2001). We need two additional ingredients. First, we recall generic properties of the concave envelope $K^+(u, U)$.

Let $(\Omega_n)_{n \in \mathbb{N}}$, be a sequence of subsets of Ω; that is, $\Omega_n \subset \Omega$ for every $n \in \mathbb{N}$. We say the sequence $(\Omega_n)_{n \in \mathbb{N}}$ increases to Ω if

$$\Omega_n \subset \Omega_{n+1} \quad \text{and} \quad \Omega = \bigcup_{n=1}^{\infty} \Omega_n.$$

Lemma 1.21 (Stability of the upper contact set)   *Let* $(\Omega_n)_{n \in \mathbb{N}}$ *be a sequence increasing to* Ω *and* $(u_n)_{n \in \mathbb{N}}$ *be a sequence of functions,* $u_n \colon \Omega_n \to \mathbb{R}$. *Suppose there exists* $u \in C(\Omega)$ *such that* $u_n \to u$ *uniformly on each* Ω_n. *Then, we have:*

(i) $\limsup_{n\to\infty} K^+(u_n, \Omega_n) \subset K^+(u, \Omega)$.
(ii) $\limsup_{n\to\infty} |K^+(u_n, \Omega_n)| \le |K^+(u, \Omega)|$.
(iii) *Denote with $K_r^+(u, U)$ the set*

$$K_r^+(u, U) = \{x \in U \mid \exists p \in B_r \text{ such that } u(y) \le u(x)$$
$$+ p \cdot (y - x) \ \forall y \in U\}.$$

Then $\limsup_{n\to\infty} K_r^+(u_n, \Omega_n) \subset K_r^+(u, \Omega)$.

Proof We start by verifying (i); let $x \in \limsup_{n\to\infty} K^+(u_n, \Omega_n)$. It means that

$$x \in \bigcap_{n=1}^{\infty} \bigcup_{k=n}^{\infty} K^+(u_n, \Omega_n);$$

hence x belongs to infinitely many sets $K^+(u_n, \Omega_n)$. As a consequence, there exists a sequence $(p_n)_{n\in\mathbb{N}}$ such that

$$u_n(y) \le u_n(x) + p_n \cdot (y - x) \tag{1.25}$$

for every $y \in \Omega_n$. For $0 < r \ll 1$, define

$$y_n := x - r\frac{p_n}{|p_n|}.$$

For sufficiently small values of $0 < r \ll 1$, we get $y_n \in \Omega_n$ and

$$u_n(y_n) \le u_n(x) - r|p_n|.$$

We also have that u_n converges uniformly to u in Ω_n. By taking limits in the previous inequality and rearranging terms we obtain

$$\inf_{z\in B_r(x)} u(z) + r \limsup_{n\to\infty} |p_n| \le u(x).$$

Hence, $|p_n|$ is bounded and p_n converges, through a subsequence if necessary, to some $p \in \mathbb{R}^d$. Using this information in (1.25) and taking limits once again we get

$$u(y) \le u(x) + p \cdot (y - x)$$

for every $y \in \Omega$, which yields (i). By noticing that

$$\liminf_{n\to\infty} |A_n| \le |\liminf_{n\to\infty} A_n|,$$

we obtain (ii) from (i). To verify (iii) it suffices to replicate the proof of (i) localizing the vector p in B_r. $\qquad\square$

In the proof of the ABP estimate we also use the notion of sup and inf convolutions. These objects are regularizations with very good properties, from the perspective of viscosity solutions. We continue with the definition and some important properties of sup and inf convolutions.

Definition 1.22 (Sup and inf convolutions)   Let $u \in C(\overline{\Omega})$. For $\varepsilon > 0$, we define the sup convolution u^ε of u by

$$u^\varepsilon(x) := \sup_{y \in \Omega} \left(u(y) - \frac{|y - x|^2}{2\varepsilon} \right).$$

Similarly, the inf convolution u_ε of u is given by

$$u_\varepsilon(x) := \inf_{y \in \Omega} \left(u(y) + \frac{|y - x|^2}{2\varepsilon} \right).$$

One notes that

$$(-u)^\varepsilon(x) = \sup_{y \in \Omega} \left(-u(y) - \frac{|y - x|^2}{2\varepsilon} \right) = - \inf_{y \in \Omega} \left(u(y) + \frac{|y - x|^2}{2\varepsilon} \right);$$

hence, $u_\varepsilon = -(-u)^\varepsilon$. Another interesting fact is that sup convolutions are always above the function u. In fact

$$u^\varepsilon(x) \geq u(y) - \frac{|y - x|^2}{2\varepsilon}$$

holds for every $y, x \in \Omega$, because of the definition of u^ε. By taking $y \equiv x$ one gets $u^\varepsilon(x) \geq u(x)$ for every $x \in \Omega$.

A more involved property concerns the set

$$A^\varepsilon(u) := \operatorname{argmax} \left(u(y) - \frac{|y - x|^2}{2\varepsilon} \right).$$

We notice $A^\varepsilon(u)$ is nonempty, because u is (uniformly) continuous up to the boundary of (the compact set) $\overline{\Omega}$. Take $x^\varepsilon \in A^\varepsilon(u)$; we have

$$u(x^\varepsilon) - \frac{|x^\varepsilon - x|^2}{2\varepsilon} \geq u(y) - \frac{|y - x|^2}{2\varepsilon},$$

for every $y \in \Omega$. By setting $y = x$, the previous inequality yields

$$\frac{|x^\varepsilon - x|^2}{2\varepsilon} \leq u(x^\varepsilon) - u(x).$$

Finally, we get

$$|x^\varepsilon - x| \leq 2 \left(\varepsilon \|u\|_{L^\infty(\Omega)} \right)^{\frac{1}{2}}. \tag{1.26}$$

The importance of (1.26) will become clear in the next lemma. For now, we observe that it quantifies a rate of convergence of x^ε to x, as $\varepsilon \to 0$. We continue with a lemma on the convergence of u^ε to u. This result combines the modulus of continuity of u with (1.26).

Lemma 1.23 (Uniform convergence of sup and inf convolutions) *Let $u \in C(\overline{\Omega})$. For $\varepsilon > 0$ let u^ε and u_ε be as in Definition 1.22. Then*

$$u^\varepsilon \longrightarrow u \qquad and \qquad u_\varepsilon \longrightarrow u$$

uniformly in Ω, as $\varepsilon \to 0$.

Proof We prove the lemma for the case of sup convolutions, the remaining one being analogous. Because u is continuous in the compact set $\overline{\Omega}$, it is uniformly continuous and has a modulus of continuity. Denote with $\omega(\cdot)$ the modulus of continuity of u in $\overline{\Omega}$. It follows that

$$
\begin{aligned}
u^\varepsilon(x) = u(x^\varepsilon) - \frac{|x^\varepsilon - x|^2}{2\varepsilon} \\
\leq u(x^\varepsilon) - u(x) + u(x) \\
\leq u(x) + \omega(|x^\varepsilon - x|).
\end{aligned}
$$

Since $u^\varepsilon \geq u$, the former inequality yields

$$|u^\varepsilon(x) - u(x)| \leq \omega\left(2\sqrt{\varepsilon\|u\|_{L^\infty(\Omega)}}\right),$$

which completes the proof.   □

Next, we state semiconcavity and semiconvexity properties for sup and inf convolutions. Those properties are instrumental in the proof of the ABP maximum principle. Once they are established, the Aleksandrov Theorem yields information on the second derivatives of u^ε and u_ε. For the sake of completeness we recall next the definition of semiconcavity and semiconvexity, as well as the Aleksandrov Theorem.

Definition 1.24 (Semiconcavity and semiconvexity) Let $v \in C(\Omega)$. We say that v is semiconcave if there exists $\varepsilon > 0$ such that

$$x \mapsto v(x) - \frac{|x|^2}{2\varepsilon}$$

is concave. We say that v is semiconvex if $-v$ is semiconcave. If v is semiconcave (semiconvex), we refer to the quantity ε as the semiconcavity (semiconvexity) constant of v

We continue with the Aleksandrov Theorem, which addresses the regularity of semiconcave and semiconvex functions. In particular, it states that a semiconcave function is twice-differentiable almost everywhere in the domain.

Proposition 1.25 (Aleksandrov Theorem) *Let $u \in C(\Omega)$ be a semiconcave function, with semiconcavity constant $\varepsilon > 0$. Then, u is twice-differentiable almost everywhere in Ω. That is, there exists a measurable function $M : \Omega \to S(d)$ such that*

$$v(y) = v(x) + Dv(x) \cdot (y - x) + \frac{1}{2}M(x)(y - x) \cdot (y - x) + o\left(|x - y|^2\right)$$

for almost every $x \in \Omega$. Moreover,

$$M(x) \geq -\frac{1}{\varepsilon}I$$

for almost every $x \in \Omega$.

The Aleksandrov Theorem is usually stated in the context of convex functions; see Niculescu and Persson (2006, Theorem 3.11.2). For a discussion, and its proof, in the case of semiconvex functions, we refer to Villani (2009, Theorem 14.1). In the former proposition, the function M stands for the Hessian of v. In fact, if we consider a mollification v_η of v, is it known that $D^2 v_\eta(x) \to M(x)$ for almost every $x \in \Omega$, as $\eta \to 0$; see Jensen (1988). In the following we prove that sup and inf convolutions are, respectively, semiconvex and semiconcave functions.

Proposition 1.26 (Semiconvexity and semiconcavity) *Let $u \in C(\overline{\Omega})$ and consider its sup and inf convolutions u^ε and u_ε given as in Definition 1.22. The sup convolution u^ε is semiconcave, whereas the inf convolution u_ε is semiconvex.*

Proof From the definition of u^ε we get

$$u^\varepsilon(x) + \frac{|x|^2}{2\varepsilon} = \sup_{y \in \Omega}\left(u(y) - \frac{|y - x|^2}{2\varepsilon}\right) + \frac{|x|^2}{2\varepsilon}$$

$$= \sup_{y \in \Omega}\left(u(y) - \frac{|y|^2}{2\varepsilon} - \frac{y \cdot x}{\varepsilon}\right).$$

We conclude that

$$u^\varepsilon(x) + \frac{|x|^2}{2\varepsilon}$$

is the supremum of the affine functions $a_y - b_y \cdot x$, where

$$a_y := u(y) - \frac{|y|^2}{2\varepsilon} \quad \text{and} \quad b_y := \frac{y}{\varepsilon}.$$

Because the supremum of affine functions is convex, we conclude that u^ε is semiconvex. The semiconcavity of u_ε follows similarly. $\qquad\square$

The last piece of information we need on sup and inf convolutions is the next lemma, which appeared in Jensen et al. (1988). For $\delta > 0$, we denote with Ω_δ the subset of Ω given by

$$\Omega_\delta := \{ x \in \Omega \,|\, \mathrm{dist}(x, \partial\Omega) > \delta \}.$$

Lemma 1.27 *Let $u \in C(\Omega)$ be a C-viscosity solution to*

$$F(x, u, Du, D^2 u) = f \quad \text{and} \quad \Omega,$$

where $F \in C(\Omega \times ,\mathbb{R} \times \mathbb{R}^d \times S(d))$ satisfies the structure condition and $f \in C(\Omega)$. For every $x^\varepsilon \in A^\varepsilon(u)$ we get

$$F\big(x^\varepsilon, u(x^\varepsilon), Du^\varepsilon(x), D^2 u^\varepsilon(x)\big) \le f(x^\varepsilon),$$

for almost every $x \in \Omega_{2(\varepsilon \|u\|_{L^\infty(\Omega)})^{1/2}}$.

For a proof of Lemma 1.27 we refer the reader to Jensen et al. (1988, Proposition 2). We notice the appearance of $\Omega_{2(\varepsilon \|u\|_{L^\infty(\Omega)})^{1/2}}$ in the previous proposition. It is necessary since $|x - x^\varepsilon|$ is bounded by $2(\varepsilon \|u\|_{L^\infty(\Omega)})^{1/2}$. Therefore, by requiring $x \in \Omega_{2(\varepsilon \|u\|_{L^\infty(\Omega)})^{1/2}}$ we ensure $x^\varepsilon \in \Omega$.

In the following, the machinery of sup and inf convolutions builds upon properties of the contact set for continuous functions to yield a version of the ABP estimate in the case $f \in C(\Omega)$.

Proposition 1.28 (The ABP estimate for continuous source terms f) *Let $u \in C(\overline{\Omega})$ be an L^d-viscosity solution to*

$$\mathcal{P}^-(D^2 u) - \gamma |Du| \le f \quad \text{in} \quad \Omega \cap \{u > 0\}, \tag{1.27}$$

where $f \in C(\Omega) \cap L^d(\Omega)$. Then there exists $C > 0$ such that

$$\sup_\Omega u \le \sup_{\partial\Omega} u^+ + C \,\mathrm{diam}(\Omega) \|f^+\|_{L^d(K^+(u))};$$

in addition, $C = C(d, \lambda, \Lambda, \mathrm{diam}(\Omega))$.

Proof We first prove the result by assuming that $u \in C^2(\Omega) \cap C(\overline{\Omega})$; then, an approximation strategy yields the result for $u \in C(\Omega)$. The proof is split into three steps; the first one gathers some technical information used further in the argument.

Step 1. Start by defining $r_0 > 0$ as

$$r_0 := \frac{\sup_\Omega u - \sup_{\partial\Omega} u^+}{\mathrm{diam}(\Omega)}.$$

For $r < r_0$, take $p \in B_r$ and denote with $x^* \in \overline{\Omega}$ a maximum point of the function $u(x) - p \cdot x$. That is,

$$u(x) \leq u(x^*) + p \cdot (x - x^*),$$

for every $x \in \overline{\Omega}$. We show next that $x^* \in \Omega$ is an interior point. In fact,

$$\sup_\Omega u(x) - u(x^*) \leq |p|\mathrm{diam}(\Omega) < r_0\mathrm{diam}(\Omega) = \sup_\Omega u - \sup_{\partial\Omega} u^+.$$

Hence $u(x^*) > \sup_{\partial\Omega} u^+$ and we conclude $x^* \in \Omega$; we also have $u(x^*) > 0$. Because $u \in C^2(\Omega)$, we know that Du exists; hence, $Du(x^*) = p$. For the same reason, D^2u exists and $D^2(u(x) - p \cdot x)$ is nonpositive at $x = x^*$. Therefore, $D^2u(x^*) \leq 0$. We also notice that $K_r^+(u)$ is a compact subset of Ω and that $D^2u \leq 0$ in $K_r^+(u)$. In addition, $u > 0$ in $K_r^+(u)$; it follows from a reasoning similar to the one yielding $u(x^*) > 0$. A more involved fact concerns the set

$$S_r := \left\{ q \in B_r \,|\, q = Du(x) \text{ for some } x \in K_r^+(u) \right\}.$$

It is clear that $B_r \subset S_r$. The converse is also true, which implies that $B_r = S_r$. At this point, we take $\kappa \geq 0$, to be fixed further in the proof, and consider a change of variables $p = Du$. Then

$$\int_{B_r} \left(|p|^{\frac{d}{d-1}} + \kappa^{\frac{d}{d-1}} \right)^{1-d} dp = \int_{K_r^+(u)} \left(|Du|^{\frac{d}{d-1}} + \kappa^{\frac{d}{d-1}} \right)^{1-d} \left| \det D^2u \right| dx$$

$$\leq \int_{K_r^+(u)} \left(|Du|^{\frac{d}{d-1}} + \kappa^{\frac{d}{d-1}} \right)^{1-d} \left(\frac{-\mathrm{Tr}\, D^2u}{d} \right)^d dx,$$

$$\tag{1.28}$$

where the last inequality follows from the fact that $M \in S(d)$, with $M \geq 0$, and

$$\det(M) \leq \left(\frac{\mathrm{Tr}\, M}{d} \right)^d.$$

In the next step, we examine the viscosity inequality satisfied by u.

Step 2. From the hypotheses of the proposition, we get

$$\mathcal{P}^-(D^2u) - \gamma|Du| \leq f(x)$$

in $\{u > 0\}$. Because $D^2u \leq 0$ in $K_r^+(u)$, the extremal operator $\mathcal{P}^-(D^2u)$ becomes

$$\mathcal{P}^-\left(D^2u\right) = -\lambda \operatorname{Tr}\left(D^2u\right)$$

in $K_r^+(u)$. Now, because $K_r^+(u) \subset \{u > 0\}$, we end up with

$$-\operatorname{Tr}\left(D^2u\right) \leq \frac{\gamma|Du| + f}{\lambda} \tag{1.29}$$

in $K_r^+(u)$. To proceed, we resort to an auxiliary inequality; for $a, b \geq 0$, we have

$$(a+b)^k \leq 2^{k-1}\left(a^k + b^k\right). \tag{1.30}$$

This inequality leads to

$$
\begin{aligned}
\left(\gamma|Du| + \frac{\kappa f^+}{\kappa}\right)^d &\leq 2^{d-1}\left(\gamma^d|Du|^d + \frac{\kappa^d\left(f^+\right)^d}{\kappa^d}\right) \\
&\leq 2^{d-1}\left(\gamma^d|Du|^d + \frac{\kappa^d\left(f^+\right)^d}{\kappa^d} + \kappa^d\gamma^d + \frac{|Du|^d\left(f^+\right)^d}{\kappa^d}\right) \\
&\leq 2^{d-1}\left(\gamma^d + \frac{\left(f^+\right)^d}{\kappa^d}\right)\left(|Du|^d + \kappa^d\right)^{\frac{d-1}{d-1}} \\
&\leq 2^{d-1}\left(\gamma^d + \frac{\left(f^+\right)^d}{\kappa^d}\right)\left(|Du|^{\frac{d}{d-1}} + \kappa^{\frac{d}{d-1}}\right)^{d-1}.
\end{aligned}
\tag{1.31}
$$

By combining (1.28), (1.29), and (1.31), we get

$$\int_{B_r}\left(|p|^{\frac{d}{d-1}} + \kappa^{\frac{d}{d-1}}\right)^{1-d}\mathrm{d}p \leq \frac{2^{d-1}}{d^d\lambda^d}\int_{K_r^+(u)}\left(\gamma^d + \frac{\left(f^+\right)^d}{\kappa^d}\right)\mathrm{d}x. \tag{1.32}$$

We resort once again to (1.30) with $a = |p|^{\frac{d}{d-1}}$, $b = \kappa^{\frac{d}{d-1}}$, and $k = d-1$; it leads to

$$2^{2-d}\frac{1}{|p|^d + \kappa^d} \leq \left(|p|^{\frac{d}{d-1}} + \kappa^{\frac{d}{d-1}}\right)^{1-d}.$$

Hence,

$$
\begin{aligned}
\frac{2^{2-d}}{d}\omega_d \ln\left(\frac{r^d}{\kappa^d} + 1\right) &= 2^{2-d}\int_{B_r}\frac{1}{|p|^d + \kappa^d}\mathrm{d}p \\
&\leq \int_{B_r}\left(|p|^{\frac{d}{d-1}} + \kappa^{\frac{d}{d-1}}\right)^{1-d}\mathrm{d}p \\
&\leq \frac{2^{d-1}}{d^d\lambda^d}\int_{K_r^+(u)}\left(\gamma^d + \frac{\left(f^+\right)^d}{\kappa^d}\right)\mathrm{d}x.
\end{aligned}
$$

Therefore,

$$\ln\left(\frac{r^d}{\kappa^d}+1\right) \le \frac{2^{2d-3}}{d^{d-1}\lambda^d\omega_d}\int_{K_r^+(u)}\left(\gamma^d+\frac{(f^+)^d}{\kappa^d}\right)dx.$$

By taking exponentials on both sides of the former inequality, and rearranging terms, we get

$$r^d \le \left(\exp\left(\frac{C(d,\lambda)}{\lambda^d}\int_{K_r^+(u)}\left(\gamma^d+\frac{(f^+)^d}{\kappa^d}\right)dx\right)-1\right)\kappa^d. \qquad (1.33)$$

At this point we choose $\kappa > 0$. Take

$$\kappa := \frac{\|f^+\|_{L^d\left(K_r^+(u)\right)}}{\lambda};$$

the inequality in (1.33) becomes

$$r \le \frac{\left(\exp\left(C(d,\lambda)\int_{K^+(u)}\frac{\gamma^d}{\lambda^d}dx+1\right)-1\right)^{\frac{1}{d}}}{\lambda}\|f^+\|_{L^d(K^+(u))}. \qquad (1.34)$$

Because

$$\sup_{\Omega}u \le \sup_{\partial\Omega}u^+ + r_0\text{diam}(\Omega),$$

we get

$$\sup_{\Omega}u \le \sup_{\partial\Omega}u^+ + C\text{diam}(\Omega)\|f^+\|_{L^d(K^+(u))},$$

where

$$C := \frac{\left(\exp\left(\frac{2^{2d-3}}{d^{d-1}\lambda^d\omega_d}\int_{K^+(u)}\frac{\gamma^d}{\lambda^d}dx+1\right)-1\right)^{\frac{1}{d}}}{\lambda}.$$

The proof is complete in the case $u \in C^2(\Omega)\cap C(\overline{\Omega})$. Next we consider merely continuous solutions and resort to the sup convolution of u.

Step 3. To remove the condition $u \in C^2(\Omega)$ we use the sup convolution u^ε. Proposition 1.26 builds upon the Alexandrov Theorem (Proposition 1.25) to ensure that u^ε is twice-differentiable almost everywhere in Ω. Because of Lemma 1.27, we have

$$\mathcal{P}^-(D^2u^\varepsilon) - \gamma|Du^\varepsilon| \le f_\varepsilon(x) \quad \text{in} \quad \Omega_{2(\varepsilon\|u\|_{L^\infty(\Omega)})^{\frac{1}{2}}},$$

where

$$f_\varepsilon(x) := \sup_{|x-y|\le 2(\varepsilon\|u\|_{L^\infty(\Omega)})^{\frac{1}{2}}} f(y).$$

Now we examine the contact sets $K_r^+(u^\varepsilon)$. Take $r < r_0$ and denote with r_0^ε the quantity

$$r_0^\varepsilon := \frac{1}{\mathrm{diam}(\Omega)} \left(\sup_\Omega u^\varepsilon - \sup_{\partial\Omega}(u^\varepsilon)^+ \right).$$

Because $u^\varepsilon \to u$ uniformly, we still have $r < r_0^\varepsilon$, provided $0 < \varepsilon \ll 1$ is taken sufficiently small. As a consequence, for every small ε, the sets $K_r^+(u^\varepsilon)$ are contained in a single compact subset of Ω.

At this point we claim that (1.28) can be written with u^ε in lieu of u. To see this is the case consider a standard mollification u_η^ε of u^ε. If $0 < \eta \ll 1$ is taken small enough, then (1.28) can be written with u_η^ε instead of u. That is,

$$
\begin{aligned}
\int_{B_r} \left(|p|^{\frac{d}{d-1}} + \kappa^{\frac{d}{d-1}} \right)^{1-d} \mathrm{d}p &= \int_{K_r^+(u_\eta^\varepsilon)} \left(|Du_\eta^\varepsilon|^{\frac{d}{d-1}} + \kappa^{\frac{d}{d-1}} \right)^{1-d} \left| \det D^2 u_\eta^\varepsilon \right| \mathrm{d}x \\
&\leq \int_{K_r^+(u_\eta^\varepsilon)} \left(|Du_\eta^\varepsilon|^{\frac{d}{d-1}} + \kappa^{\frac{d}{d-1}} \right)^{1-d} \left(\frac{-\operatorname{Tr} D^2 u_\eta^\varepsilon}{d} \right)^d \mathrm{d}x.
\end{aligned}
$$

$$(1.35)$$

Our goal is to take the limit $\eta \to 0$ in (1.35). In fact, $u_\eta^\varepsilon : \Omega_\eta \to \mathbb{R}$, where the sets $\Omega_{\eta_0} \subset \Omega_{\eta_1}$ if $\eta_0 \geq \eta_1$. Moreover,

$$\Omega = \bigcup_{\eta > 0} \Omega_\eta.$$

Hence, the family $\{\Omega_\eta\}_{\eta > 0}$ exhausts Ω. Because of (iii) in Lemma 1.21, we get

$$\limsup_{\eta \to 0} K_r^+\left(u_\eta^\varepsilon \right) \subset K_r^+\left(u^\varepsilon \right).$$

Moreover, we recall that $D^2 u_\eta^\varepsilon$ converges to $D^2 u^\varepsilon$ almost everywhere in Ω; we also have

$$-\frac{1}{\varepsilon} I \leq D^2 u_\eta^\varepsilon \leq 0$$

in $K_r^+(u_\eta^\varepsilon)$. As a result, we can take the limit $\eta \to 0$ in (1.35) to obtain

$$\int_{B_r} \left(|p|^{\frac{d}{d-1}} + \kappa^{\frac{d}{d-1}} \right)^{1-d} \mathrm{d}p \leq \int_{K_r^+(u^\varepsilon)} \left(|Du^\varepsilon|^{\frac{d}{d-1}} + \kappa^{\frac{d}{d-1}} \right)^{1-d} \left(\frac{-\operatorname{Tr} D^2 u^\varepsilon}{d} \right)^d \mathrm{d}x.$$

$$(1.36)$$

The computations in Step 2 remain unchanged and we get

$$r \leq \frac{\left(\exp\left(C(d,\lambda) \int_{K^+(u^\varepsilon)} \frac{\gamma^d}{\lambda^d} \mathrm{d}x + 1 \right) - 1 \right)^{\frac{1}{d}}}{\lambda} \left\| (f^\varepsilon)^+ \right\|_{L^d(K^+(u^\varepsilon))}. \quad (1.37)$$

We take the limit $\varepsilon \to 0$ in (1.37); item (iii) in Lemma 1.21 and the continuity of f yield (1.34) for the merely continuous function u, and the proof is complete.   □

Now we are in a position to prove Proposition 1.20. The strategy is to consider a sequence $(f_n)_{n \in \mathbb{N}} \subset L^d(\Omega) \cap C^\infty(\Omega)$ converging to $f \in L^d(\Omega)$ in the L^d-topology.

Proof of Proposition 1.20   Let $(f_n)_{n \in \mathbb{N}} \subset L^d(\Omega) \cap C^\infty(\Omega)$ be such that $f_n \to f$ in $L^d(\Omega)$. Because of Proposition 1.16, we know there exists a strong solution $\varphi_n \in W^{2,d}_{\mathrm{loc}} \cap C(\overline{\Omega})$ to

$$\mathcal{P}^+\left(D^2\varphi_n\right) + \gamma\left|D\varphi_n\right| \leq f_n - f \quad \text{in} \quad \Omega,$$

with $\varphi_n = 0$ on $\partial\Omega$. We learn from the estimate in (1.15) that $\|\varphi\|_{L^\infty(\Omega)} \to 0$ as $n \to \infty$.

Now we consider an auxiliary function. Let $w := u + \varphi_n - \|\varphi\|_{L^\infty(\Omega)}$. We get

$$\mathcal{P}^-\left(D^2w\right) - \gamma|Dw| \leq \mathcal{P}^-\left(D^2u\right) - \gamma|Du| + \mathcal{P}^+\left(D^2\varphi_n\right) + \gamma\left|D\varphi_n\right|.$$

Since $\varphi \leq \|\varphi\|_{L^\infty(\Omega)}$, it follows that $\{w > 0\} \subset \{u > 0\}$. Hence,

$$\mathcal{P}^-\left(D^2w\right) - \gamma|Dw| \leq f_n \quad \text{in} \quad \Omega \cap \{w > 0\}.$$

Because f_n is continuous, we infer from Proposition 1.28 that

$$\sup_\Omega w \leq \sup_{\partial\Omega} w^+ + C\mathrm{diam}(\Omega)\left\|f_n\right\|_{L^d(K^+(w))};$$

by taking the limit $n \to \infty$ in the former inequality we complete the proof.   □

1.2 Krylov–Safonov Theory

In this section we detail a fundamental result in the study of elliptic equations in nondivergence form, due to Krylov and Safonov (1980). It comprises a Harnack inequality and C^α-regularity estimates. To briefly add some context, it is worth mentioning that the Krylov–Safonov theory accounts for the nonvariational counterpart of the celebrated theory of De Giorgi–Nash–Moser theory, available for elliptic equations in divergence form.

Another interesting aspect of the theory is that it applies to subsolutions, or supersolutions, to an elliptic equation. In other words, the Krylov–Safonov estimates are available for functions satisfying an *inequality* instead of an equality.

In what follows, we consider L^p-viscosity solutions to

$$\mathcal{P}^+_{\lambda,\Lambda}\left(D^2 u\right) \geq 0 \geq \mathcal{P}^-_{\lambda,\Lambda}\left(D^2 u\right) \quad \text{in} \quad \Omega$$

and prove an weak L^ε-estimate of the form

$$\left|\left\{ u > 2^{C(\lambda,\Lambda,d)} u(0) t \right\}\right| \leq C t^{-\varepsilon}. \tag{1.38}$$

Our approach follows very closely the strategy in Mooney (2019). The general idea underlying the argument is to touch the graph of solutions with paraboloids of certain openings. Given a fixed number $a \geq 0$, we consider paraboloids of opening a and collects all the points where the graph of u can be touched by a paraboloid of such opening. Then we examine the measure decay of those sets, as the opening a increases – in a dyadic scale, for example. This is a powerful idea, which we revisit in Chapter 2, Section 3.2, and Section 4.3 in different contexts.

The main theorem in this section is phrased in terms of a regularity result. It reads as follows.

Theorem 1.29 (Krylov–Safonov regularity estimate)   *Let $u \in C(\Omega)$ be a viscosity solution to*

$$\mathcal{P}^+_{\lambda,\Lambda}\left(D^2 u\right) \geq 0 \geq \mathcal{P}^-_{\lambda,\Lambda}\left(D^2 u\right) \quad \text{in} \quad \Omega$$

Then $u \in C^\alpha_{loc}(\Omega)$ for some $\alpha \in (0,1)$, with $\alpha = \alpha(d,\lambda,\Lambda,\mathrm{diam}(\Omega))$. In addition, for every $\Omega' \Subset \Omega$, we have the estimate

$$\|u\|_{C^\alpha(\Omega')} \leq C \|u\|_{L^\infty(\Omega)},$$

where $C = C(d,\lambda,\Lambda,\mathrm{diam}(\Omega),\mathrm{dist}(\Omega',\partial\Omega))$ is a nonnegative constant.

We present the proof recently developed by Mooney (2019). Its main idea is to prove an L^ε-estimate; to that end, the author touches the graph of the solutions with paraboloids and produces a convex envelope and a contact set. Then one takes a paraboloid of larger opening touching the previous contact set; as the paraboloid is slid up until it touches the solution, a new contact point arises. The proof relies on the comparison between the contact points at the two levels.

Once an appropriate oscillation control is available, we resort to a Gronwall-type of inequality; see, for instance, Gilbarg and Trudinger (2001, Theorem 8.23). For the sake of completeness we recall it next.

Lemma 1.30   *Let $R > 0$ be fixed and consider $\omega: [0,R] \rightarrow \mathbb{R}$. Suppose ω is nondecreasing; suppose further there exist constants $\gamma, \tau \in (0,1)$ and a nondecreasing function $\sigma: \mathbb{R} \rightarrow \mathbb{R}$ satisfying*

$$\omega(\tau r) \le \gamma \omega(r) + \sigma(r)$$

for every $r \in [0, R]$. *Then, for every* $\mu \in (0, 1)$ *there exist* $\alpha = \alpha(\gamma, \tau, \mu) \in (0, 1)$ *and a positive constant* $C = C(\gamma, \tau, \mu)$ *such that*

$$\omega(r) \le C \left(\left(\frac{r}{R} \right)^{\alpha} \omega(R) + \sigma \left(r^{\mu} R^{1-\mu} \right) \right).$$

The proof of Lemma 1.30 can be found in Gilbarg and Trudinger (2001, p.~201) and we omit it here. In the following, we introduce the main structures we use in this section. The paraboloid $P_{y,b}^A(x)$ is the function defined by

$$P_{y,b}^A(x) := b + y \cdot x + \frac{A}{2} |x|^2,$$

where $A, b \in \mathbb{R}$ and $y \in \mathbb{R}^d$ are fixed. We say that $P_{y,b}^A$ has opening A. We are interested in the collection of paraboloids touching subsolutions from below; therefore, we focus on concave paraboloids $P_{y,b}^{-A}$, for $A > 0$.

At this point we impose a further condition on the domain $\Omega \subset \mathbb{R}^d$; in addition to requiring it to be open and bounded, we also ask it to be strictly convex. For the sake of completeness we add the following definition.

Definition 1.31 (Strictly convex sets) Let $\Omega \subset \mathbb{R}^d$. We say Ω is strictly convex if, for every $x, y \in \Omega$ and every $\tau \in (0, 1)$, we have $\tau x + (1 - \tau) y \in$ int Ω. In other terms,

$$\{\tau x + (1 - \tau) y \,|\, \text{for every } \tau \in (0, 1) \text{ and } x, y \in \Omega\} \subset \text{int } \Omega.$$

A set is strictly convex if the segment connecting any two points in it lies in the interior of the set, except for the extremal points. For example, in $\mathbb{R}^2$ consider the closed unit ball in the uniform topology – also known as a square of side length 2, centered at the origin. This set is clearly convex; however, any two adjacent vertices are connected by the side of the square, which is not in its interior. Therefore, such set is not strictly convex.

Definition 1.32 (A-convex envelope and A-contact set) Let $\Omega \subset \mathbb{R}^d$ be a strictly convex domain and take $v \in C(\overline{\Omega})$. The A-convex envelope of v in $\overline{\Omega}$ is the function $\Gamma_v^A : \overline{\Omega} \to \mathbb{R}$ defined by

$$\Gamma_v^A(x) := \sup_{\substack{y \in \mathbb{R}^d \\ b \in \mathbb{R}}} \left\{ P_{y,b}^{-A} \,|\, P_{y,b}^{-A}(x) \le v(x), \text{ for every } x \in \overline{\Omega} \right\}.$$

The A-contact set of v in Ω is denoted by K_A and given by

$$K_A(v) := \left\{ x \in \Omega \,|\, v(x) = \Gamma_v^A(x) \right\}.$$

We say that v has a tangent paraboloid of opening A at $x_0 \in \Omega$ if there exists $P_{y,b}^{-A}$ such that

$$P_{y,b}^{-A}(x) \le v(x) \qquad \text{in} \qquad \overline{\Omega},$$

with $P_{y,b}^{-A}(x_0) = v(x_0)$. When defining the A-convex envelope of v we considered the points $x \in \overline{\Omega}$, whereas in the definition of the A-contact set we restricted our attention to $x \in \Omega$. In fact, the strict convexity of the domain implies that v agrees with Γ_v^A on $\partial\Omega$.

Because Γ_v^A is convex, it is continuous in Ω; hence, $K_A(v)$ is the preimage of $\{0\}$ by a continuous function and therefore closed in Ω. Also, if $A \le \overline{A}$ then $K_A(v) \subset K_{\overline{A}}(v)$. We observe that Γ_v^0 recovers the usual convex envelope, and A_0 amounts to the associated contact set. Before proceeding we notice two other facts.

For $\lambda, \gamma \in \mathbb{R}$ and $\mu > 0$ we claim that

$$\Gamma^{\lambda}_{\mu v + \frac{\gamma}{2}|x|^2} = \mu \Gamma_v^{\frac{\lambda+\gamma}{\mu}} + \frac{\gamma}{2}|x|^2 \quad \text{and} \quad K_\lambda\left(\mu v + \frac{\gamma}{2}|x|^2\right) = K_{\frac{\lambda+\gamma}{\mu}}(v). \tag{1.39}$$

For the first equality, start by noticing that

$$P_{y,b}^{-\lambda} \le \mu v + \frac{\gamma}{2}|x|^2,$$

if and only if

$$\frac{P_{y,b}^{-\lambda} - \frac{\gamma}{2}|x|^2}{\mu} \le v.$$

Hence we are interested in the quantity

$$\sup_{\substack{y \in \mathbb{R}^d \\ b \in \mathbb{R}}} \left\{ P_{y,b}^{-\lambda}(x) \mid P_{y/\mu,b/\mu}^{\frac{-\lambda-\gamma}{\mu}}(x) \le v(x) \right\},$$

which leads to the equality. The second equality, concerning the contact sets, is then immediate. We proceed with three auxiliary lemmas; see Mooney (2019, Lemmas 3.1–3.3).

Lemma 1.33 *Let $u \in C(\overline{\Omega})$ be an L^p-viscosity solution to*

$$\mathcal{P}_{\lambda,\Lambda}^+(D^2 u) \ge C \quad \text{in} \quad \Omega. \tag{1.40}$$

For $A > 0$, suppose there exists a paraboloid $P_{y,b}^{-A}$ such that

$$P_{y,b}^{-A} \le u \text{ in } \overline{\Omega} \quad \text{but} \quad P_{y,b}^{-A}(x_0) = u(x_0),$$

for some $x_0 \in \Omega \setminus K_0(v)$. Slide up $P_{y,b}^{-A}$ by considering $P_{y,b+t}^{-A}$ until it touches u from below at $x_1 \in \overline{\Omega}$, for some $t > 0$. Then, $x_1 \in \overline{\Omega} \setminus K_0(u)$.

Proof The proof follows from a contradiction argument; we suppose $x_1 \in K_0(u)$ and consider two cases. In fact, if $x_1 \in K_0(u)$, there exists a supporting hyperplane ℓ to Γ_u^0 at x_1. Then two possibilities arise: either $P_{y,b+t}^{-A}$ is tangent from below to ℓ or not.

Case 1. Suppose first that $P_{y,b+t}^{-A}$ is tangent from below to the supporting hyperplane ℓ. In this case,

$$P_{y,b+t}^{-A}(x) \le \ell(x) \le \Gamma_u^0(x) \quad \text{in} \quad \Omega.$$

Because $P_{y,b}^{-A} < P_{y,b+t}^{-A}$, the former inequality yields $P_{y,b}^{-A}(x_0) < \Gamma_u^0$, which is a contradiction. It remains for us to examine the case where $P_{y,b+t}^{-A}$ is not tangent from below to ℓ.

Case 2. Suppose otherwise that $P_{y,b+t}^{-A}$ is not tangent from below to ℓ. Subtract ℓ from u and resort to the usual transformations (horizontal translations, rotations, rescaling, and multiplications by a constant) to ensure that $e_1 \in \Omega$ and $\max(0, 1 - |x|^2)$ is tangent from below to u at e_1.

Now, for $a > 1$, we consider $\varphi_a : \Omega \to \mathbb{R}$ given by

$$\varphi_a(x) := 1 - |x| + a(|x| - 1)^2;$$

notice it is of class C^∞ in a vicinity of e_1. For

$$x \in \left[\frac{a}{1+a}, \frac{1+a}{a} \right],$$

we have $\varphi_a(x) \le \max(0, 1 - |x|^2)$. Hence, φ_a is tangent to u from below at e_1; now we use the equation and the notion of viscosity solution. Because u is a viscosity solution to (1.40), we have

$$\mathcal{P}_{\lambda,\Lambda}^+ \left(D^2 \varphi_a(e_1) \right) \ge C. \tag{1.41}$$

On the other hand, $D^2 \varphi_a(x) = (a_{i,j}(x))_{i,j=1}^d$, where

$$a_{i,j}(x) = \left(2a + \frac{1}{|x|} - \frac{x_i x_j}{|x|^3} \right) \delta_{i,j} + \left(-\frac{x_i x_j}{|x|^3} \right) (1 - \delta_{i,j}),$$

for $i, j = 1, \ldots, d$, where $\delta_{i,j} = 1$ if $i = j$ and $\delta_{i,j} = 0$ otherwise. As a consequence,

$$\mathcal{P}_{\lambda,\Lambda}^+ \left(D^2 \varphi_a(e_1) \right) = -2a\lambda + (d-1)\Lambda;$$

by choosing $a \gg 1$ large enough, we get a contradiction with (1.41) and complete the proof. $\qquad\square$

We continue with a measure-theoretical lemma relating the set of vertices of the tangent paraboloids with the set of their contact points.

Lemma 1.34 *Let $v \in C(\overline{\Omega})$ be convex and $F \subset \Omega$ be a measurable set. Denote by V the set of vertices of all paraboloids of opening $-A < 0$ tangent from below to v at points in F. Then $|V| \geq |F|$.*

Proof We start with elementary computations. The vertex of $P_{y,b}^{-A}$ is the point $x = (1/A)y$. Moreover, if $P_{y,b}^{-A}$ is tangent from below to v at some point x^*, we have

$$-Ax^* + y = p \in \partial^- v(x^*),$$

where $\partial^- g(x)$ denotes the subdifferential of the convex function g at the point x. By multiplying the former equality by $1/A$ and rearranging the terms we learn the vertices in V are of the form

$$x^* + \frac{1}{A}p,$$

where $x^* \in F$; if Dv exists at x^*, we can replace p in the former inequality with $Dv(x^*)$.

Because Dv exists for almost every $x \in \Omega$, we know that $|F| = |\overline{F}|$, where $\overline{F} := \{x \in F \mid Dv \text{ exists at } x\}$. If we denote by $w(\cdot)$ the function

$$w(x) := \frac{1}{A}v(x) + \frac{1}{2}|x|^2,$$

we notice that $Dw(\overline{F}) = \overline{V} \subset V$, and $|V| = |\overline{V}|$. We proceed by showing that $|\overline{V}| \geq |\overline{F}|$.

Indeed, w is a semiconvex function; as such, it is twice-differentiable in the sense of Proposition 1.25. Moreover,

$$D^2 w(x) = I + \frac{1}{A}D^2 v(x) \geq I,$$

where the second inequality follows from the convexity of v. As a conclusion, $\det D^2 w(x) \geq 1$; an application of the change of variables formula yields

$$|\overline{V}| = \int_V \mathrm{d}x = \int_F |\det D^2 w| \, \mathrm{d}x \geq |F|,$$

which completes the proof.   □

In what follows, we use the former lemmas to prove a measure estimate. First, we prove an intermediate lemma, under the assumption that the solution u is semiconvex. Then a second lemma uses sup and inf convolutions to conclude the measure estimate for the general case.

Lemma 1.35 (Measure estimate for semiconvex solutions) *Let $u \in C(\overline{\Omega})$ be a viscosity solution to*

$$\mathcal{P}^+(D^2 u) \geq 0 \quad in \quad \Omega.$$

Suppose u is semiconvex. For $A > 0$ and a measurable set $F \subset \{u > \Gamma_u^A\}$ consider the paraboloids of opening $-2A$ tangent to Γ_u^A on the set F. Slide those paraboloids up until they touch u; denote with E the set of points where the slid paraboloids touch u. If $E \Subset \Omega$, then

$$|K_{2A}(u) \setminus K_A(u)| \geq 2^{-1}(2d\Lambda/\lambda)^{1-d}|F|.$$

Proof The proof is split into two steps. Because F is measurable, for every $\delta > 0$ there exists a closed subset $\overline{F} \subset F$ such that $|F \setminus \overline{F}| < \delta$. Therefore, we suppose from the beginning that F is closed. We start by verifying the inclusion $E \subset K_{2A}(u) \setminus K_A(u)$.

Step 1. Consider the auxiliary function

$$v(x) := \frac{1}{A}u(x) + \frac{1}{2}|x|^2;$$

we write $K_{2A}(u)$ and $K_A(u)$ in terms of the contact sets for v. Because of (1.39) we have

$$K_{2A}(u) = K_1(v) \quad and \quad K_A(u) = K_0(v).$$

In this rescaled setting, take the paraboloids of opening -1 which are tangent from below to Γ_v^0 at points in F; slide them up until they touch v. The set E now comprises the points those paraboloids touch v; clearly E is closed. Denote by V the (closed) set of the vertices of those paraboloids; Lemma 1.34 yields $|F| \leq |V|$.

We compute $\mathcal{P}^+(D^2 v)$ to find

$$0 \leq \frac{1}{A}\mathcal{P}^+(D^2 u) \leq \mathcal{P}^+(D^2 v) + \Lambda d.$$

Hence, $\mathcal{P}^+(D^2 v) \geq -\Lambda d$ and v falls within the scope of Lemma 1.33. Notice that if $x^* \in E$ it follows from Lemma 1.33 that $x^* \in \Omega \setminus K_0(v)$ and $E \cap K_0(v) = \emptyset$. On the other hand, $E \subset K_1(v)$ and we get $E \subset K_1(v) \setminus K_0(v)$ and, immediately, $|E| \leq |K_1(v) \setminus K_0(v)|$. In the next step we control the measure of V is terms of the measure of E.

Step 2. Because u is semiconvex, we conclude that v is twice-differentiable almost everywhere in Ω. Moreover, at a twice-differentiability point x, we have

$$\mathcal{P}^+(D^2 v(x)) \geq -\Lambda d; \tag{1.42}$$

see, for instance, Caffarelli et al. (1996, Proposition 3.4).

If $x \in E$, there exists a paraboloid $P_{y,b}^{-1}$ touching v at x^*; in addition, the semiconvexity assumption implies that v is differentiable. Hence,

$$y = x^* + Dv(x^*).$$

The vertex of $P_{y,b}^{-1}$ is located at y; hence $x \mapsto x + Dv(x)$ maps E into V. This map is Lipschitz continuous in E; in fact, for $x_1, x_2 \in E$, we have

$$|x_1 + Dv(x_1) - x_2 - Dv(x_2)| \leq |x_1 - x_2| + |-x_1 + y - x_2 - y| \leq 2|x_1 - x_2|.$$

Now, in the case where x is a point of twice-differentiability of v, we find $D^2 v \geq -I$; it follows from the fact that $v - P_{y,b}^{-1}$ has a minimum point at x and its Hessian is nonnegative. We combine this information with (1.42) to produce

$$-\Lambda d \leq \mathcal{P}^+ \left(D^2 u \right) \leq \mathcal{P}^+ \left(D^2 v + I - I \right) - \mathcal{P}^-(I)$$
$$\leq -\lambda \operatorname{Tr} \left(D^2 v \right) - \lambda d + \Lambda d.$$

Rearranging the terms we get

$$\operatorname{Tr} \left(D^2 v \right) \leq \left(2 \frac{\Lambda}{\lambda} - 1 \right) \operatorname{Tr}(I).$$

Hence,

$$I + D^2 v \leq \left(2 \frac{\Lambda}{\lambda} \right) I.$$

Using the equation for u, we infer that the smallest eigenvalue of $D^2 u$ is nonpositive. This fact implies that the smallest eigenvalue of $D^2 v$ is smaller than or equal to 1. As a result,

$$\det(I + D^2 v) \leq 2 \left(2 \frac{\Lambda}{\lambda} \right)^{d-1}.$$

The key point in the inequality above is that it produces an upper bound for the Jacobian associated with the map $E \to V$. Therefore,

$$|V| = \int_V \mathrm{d}x \leq 2 \left(2 \frac{\Lambda}{\lambda} \right)^{d-1} \int_E \mathrm{d}x = 2 \left(2 \frac{\Lambda}{\lambda} \right)^{d-1} |E| \leq 2 \left(2 \frac{\Lambda}{\lambda} \right)^{d-1} |K_1(v) \setminus K_0(v)|.$$

Lemma 1.33 yields $|F| \leq |V|$ and the proof is complete. □

Now we state and prove the measure estimate for general u, removing the assumption of semiconvexity.

Lemma 1.36 (Measure estimate)   *Let $u \in C(\overline{\Omega})$ be a viscosity solution to*

$$\mathcal{P}^+ \left(D^2 u \right) \geq 0 \quad in \quad \Omega.$$

For $A > 0$ and a measurable set $F \subset \{u > \Gamma_u^A\}$ consider the paraboloids of opening $-2A$ tangent to Γ_u^A on the set F. Slide those paraboloids up until they touch u; denote by E the set of points where the slid paraboloids touch u. If $E \Subset \Omega$, then

$$|K_{2A}(u) \setminus K_A(u)| \geq 2^{-1}(2d\Lambda/\lambda)^{1-d}|F|.$$

Proof If $E \Subset \Omega$, we take Ω' such that $E \Subset \Omega' \Subset \Omega$. Consider the family of sup convolutions u^ε, given as in Definition 1.22. It is known that u^ε converges uniformly to u in Ω', as $\varepsilon \to 0$. Moreover,

$$\mathcal{P}^+(D^2 u^\varepsilon) \geq 0 \quad \text{in} \quad \Omega_{2(\varepsilon\|u\|_{L^\infty(\Omega)})^{\frac{1}{2}}}.$$

Define $v^\varepsilon(\cdot)$ as

$$v^\varepsilon(x) := \frac{1}{A} u^\varepsilon(x) + \frac{1}{2}|x|^2,$$

and consider the associated sets E^ε. As before,

$$\mathcal{P}^+(D^2 v^\varepsilon) \geq -\Lambda d$$

and, for $0 < \varepsilon \ll 1$ sufficiently small, $E^\varepsilon \Subset \Omega'$. Arguing as in the proof of Lemma 1.35 we get

$$|F| \leq |V| \leq 2 \left(2\frac{\Lambda}{\lambda}\right)^{d-1} |E^\varepsilon|.$$

Because

$$\limsup_{\varepsilon \to 0} E^\varepsilon \subset E \subset (K_1(v) \setminus K_0(v)),$$

the result follows. $\square$

In the following we use the measure estimate to prove an L^ε-estimate; we continue with a proposition.

Proposition 1.37 (Oscillation estimate)   *Let $u \in C(\Omega)$ be a nonnegative viscosity solution to*

$$\mathcal{P}^+(D^2 u) \geq 0 \quad \text{in} \quad B_{4\sqrt{d}}.$$

Then

$$\inf_{x \in B_\rho(x_0)} u(x) \leq 2\rho^{-d\Lambda/\lambda} u(0),$$

for every $x_0 \in B_1$ and $0 < \rho < 1/2$ such that $B_\rho(x_0) \subset B_1$.

Proof We examine two possibilities. First, suppose $0 \in \overline{B_\rho(x_0)}$. In this case

$$\inf_{x \in B_\rho(x_0)} u(x) \le u(0) < 2\rho^{d\Lambda/\lambda} u(0),$$

and the result follows. Now we consider the case $0 \notin \overline{B_\rho(x_0)}$ and resort to a contradiction argument.

Suppose that

$$\inf_{x \in B_\rho(x_0)} u(x) > 2\rho^{d\Lambda/\lambda} u(0).$$

For $x_0 \in B_1$ fixed, define the auxiliary function w by

$$w(x) := \frac{|x - x_0|^{-d\Lambda/\lambda} - 2^{-d\Lambda/\lambda}}{1 - 2^{-d\Lambda/\lambda}}.$$

In $B_4 \setminus \overline{B_\rho(x_0)}$ we compute the Hessian of w to obtain

$$\frac{\partial^2}{\partial x_i \partial x_j} w(x) = C(d, \Lambda, \lambda) \frac{(x_i - x_0)(x_j - x_0)}{|x - x_0|^{d\frac{\Lambda}{\lambda}+4}} - \delta_{i,j} \frac{d\Lambda}{\lambda |x - x_0|^{d\frac{\Lambda}{\lambda}+2}},$$

where

$$C(d, \lambda, \Lambda) := d\frac{\Lambda}{\lambda} \left(d\frac{\Lambda}{\lambda} + 2 \right).$$

Because $u(0) \ge 0$, we infer that

$$\mathcal{P}^+\left(D^2(u(0)w)\right) < 0 \quad \text{in} \quad B_4 \setminus \overline{B_\rho(x_0)};$$

also, because $w < 0$ on ∂B_4, we have $u(0)w < u$ on ∂B_4. Moreover, since $w(0) > 1$, we get

$$u(0) < u(0)w(0); \tag{1.43}$$

that is, at the interior point $x = 0$ we have $u > u(0)w$. Finally, notice that $w < 2\rho^{-d\Lambda/\lambda}$ on $\partial B_\rho(x_0)$. Were $u > 2\rho^{-d\Lambda/\lambda} u(0)$ on $\partial B_\rho(x_0)$, we would have $u > u(0)w$ on $\partial B_\rho(x_0)$. In brief,

$$\begin{cases} \mathcal{P}^+\left(D^2(u(0)w)\right) < \mathcal{P}^+\left(D^2 u\right) & \text{in} \quad B_4 \setminus \overline{B_\rho(x_0)}, \\ u(0)w < u & \text{on} \quad \partial B_4 \cup \partial B_\rho(x_0). \end{cases}$$

The maximum principle yields a contradiction with (1.43) and we conclude $u \le 2\rho^{-d\Lambda/\lambda} u(0)$ on $\partial B_\rho(x_0)$; a further application of the maximum principle completes the proof. $\square$

Before continuing, we detail a useful inequality.

Lemma 1.38 (A useful inequality) *Let $d \geq 2$ and $0 < \lambda \leq \Lambda$; we have*

$$2^{-\frac{\lambda}{2d\Lambda}} < 1 - 2^{-2}(2d\Lambda/\lambda)^{1-d}. \tag{1.44}$$

Proof Set $\varphi(x) := 2^x$ and notice that $\varphi'(x) = (\ln 2)2^x$ is increasing. Observe also that

$$\varphi'\left(-\frac{1}{4}\right) = (\ln 2)2^{-\frac{1}{4}} > \frac{1}{4}.$$

For $z \in (0, 1/4)$ and $d > 2$ we get

$$\frac{1}{4}z < z \cdot \varphi'\left(-\frac{1}{4}\right) < z \cdot \varphi'(-z) < \varphi(0) - \varphi(-z) = 1 - \varphi(-z).$$

Hence,

$$\left(\frac{1}{2}\right)^z < 1 - \frac{1}{4}z < 1 - \frac{1}{4}z^{d-1},$$

since $d \geq 2$. Take

$$z := \frac{\lambda}{2d\Lambda} \in \left(0, \frac{1}{4}\right)$$

to conclude

$$\left(\frac{1}{2}\right)^{\frac{\lambda}{2d\Lambda}} < 1 - \frac{1}{4}\left(\frac{\lambda}{2d\Lambda}\right)^{d-1}$$

and complete the proof.   □

Our next proposition is an L^ε-estimate.

Proposition 1.39 (L^ε-estimate) *Let $u \in C(B_4)$ be a nonnegative viscosity solution to*

$$\mathcal{P}^+(D^2u) \geq 0 \quad in \quad B_4.$$

Then there exists $C > 0$ and $\varepsilon > 0$ such that

$$\left|\{u \geq C(d, \lambda, \Lambda)u(0)t\} \cap B_{1/2}\right| \leq |B_{1/2}|t^{-\varepsilon},$$

for every $t \geq 2$, with $\varepsilon = \varepsilon(d, \lambda, \Lambda)$ and $C = C(d, \lambda, \Lambda)$.

Proof We start with a simplification. Divide u by $2^N u(0)$, for some $N > 0$. Hence, $u(0) = 2^{-N}$. For ease of presentation, we split the proof into four steps.

Step 1. We argue by contradiction to verify that

$$K_{2^n}(u) \cap B_{1/2} \subset \{u \leq 2^n\},$$

for every $n \in \mathbb{N}$. Suppose there exists $x_0 \in K_{2^n}(u) \cap B_{1/2}$ such that $u(x_0) > 2^k$. Then, there exists a ball B, of radius 1 containing x_0, such that $u \geq 2^{n-1}$ in B. As a consequence, $u(x) \geq 2^{n-1}$ for every $x \in B_{1/4}(x_1)$, for some $x_1 \in B_1$.

From Proposition 1.37 we infer

$$\inf_{x \in B_{1/4}(x_1)} u(x) \leq 2^{1+2d\Lambda/\lambda - N} \leq 2^{-1},$$

which is a contradiction. Also, if we set

$$\rho_n := 2^{-30d - \frac{(n+1)\lambda}{2d\Lambda}},$$

then

$$\inf_{y \in B_{\rho_n}(x)} \left(2^{-n} u(y)\right) \leq \rho_n^2, \tag{1.45}$$

for every $x \in B_1$. To verify (1.45) we use once again Proposition 1.37; notice that

$$\rho_n^2 \inf_{y \in B_{\rho_n}(x)} \left(2^{-n} u(y)\right) \leq 2^{2+60d-34d^2} 2^{(n+1)\left(\frac{\lambda}{d\Lambda} - \frac{1}{2}\right)} \leq 2^{2+60d-34d^2},$$

where the second inequality follows from

$$\frac{\lambda}{d\Lambda} - \frac{1}{2} < 0.$$

Finally, for $d > 2$, we have $2^{60d-34d^2} \leq 2^{-8d}$. Hence,

$$\rho_n^2 \inf_{y \in B_{\rho_n}(x)} \left(2^{-n} u(y)\right) \leq 2^{2-8d} < 1,$$

which yields (1.45).

Step 2. Next, we use an induction argument to prove that

$$\left| B_{1/2} \setminus K_{2^n}(u) \right| \leq \left(1 - 2^{-2}(2d\Lambda/\lambda)^{1-d}\right)^n \left| B_{1/2} \right|, \tag{1.46}$$

for every $n \in \mathbb{N}$. The case $n = 0$ is immediate. Suppose the case $n = k$ has already been verified. We examine the case $n = k + 1$. In the following we prove a claim used in the induction argument; it allows us to apply Lemma 1.36.

Start with a paraboloid of opening -2, denoted by P^{-2}. Suppose P^{-2} touches $\Gamma^1_{2^{-n}}$ from below at $x_0 \in B_{1/2-5\rho_n}$ and slide it up until it touches the graph of $2^{-n}u$ at a point denoted with x_1. We claim that $x_1 \in B_{1/2-\rho_n}$.

Indeed, we have $0 \leq \Gamma^1_{2^{-n}u} \leq 2^{-n}u$. Let y denote the vertex of P^{-2} and suppose $|y - x_0| > 2\rho_n$. Then $2^{-k}u \geq 3\rho_n^2$ in $B_{\rho_n}(y)$, which is impossible because of (1.45). Hence $|y - x_0| \leq 2\rho_n$. Similarly, $|x_1 - y| \leq 2\rho_n$, which proves the claim.

Step 3. If we set $F_n := B_{1/2-5\rho_n} \setminus K_{2^n}(u)$ and let E_n denote the set of new contact points, Lemma 1.36 ensures that

$$\left| \left(K_{2^{n+1}}(u) \setminus K_n(u) \right) \cap B_{1/2} \right| \geq 2^{-1} (2d\Lambda/\lambda)^{1-d} |F_n| . \tag{1.47}$$

Now we consider two cases. Suppose first that $|F_k| < \left| B_{1/2} \setminus K_{2^{k+1}}(u) \right| /2$. This does not depend on the induction hypothesis; alternatively, it uses the inequality (1.44). Indeed, because $|F_k| < \left| B_{1/2} \setminus K_{2^{k+1}}(u) \right| /2$, we get

$$\left| B_{1/2} \setminus K_{2^{k+1}}(u) \right| \leq \left| B_{1/2} \setminus K_{2^k}(u) \right| \leq \left| B_{1/2} \setminus B_{1/2-5\rho_k} \right| \leq 2^{-\frac{(k+1)\lambda}{2d\Lambda}} \left| B_{1/2} \right| ;$$

hence, (1.44) yields

$$\left| B_{1/2} \setminus K_{2^{k+1}}(u) \right| \leq \left(1 - 2^{-2} (2d\Lambda/\lambda)^{1-d} \right)^{k+1} \left| B_{1/2} \right| .$$

Now we treat the remaining case: $|F_k| \geq \left| B_{1/2} \setminus K_{2^{k+1}}(u) \right| /2$. Notice that

$$\left| B_{1/2} \setminus K_{2^{k+1}}(u) \right| = \left| B_{1/2} \setminus K_{2^k}(u) \right| - \left| \left(K_{2^{n+1}}(u) \setminus K_n(u) \right) \cap B_{1/2} \right| .$$

Combining the previous identity with (1.47) one obtains

$$\left| B_{1/2} \setminus K_{2^{k+1}}(u) \right| \leq \left(1 - 2^{-2} (2d\Lambda/\lambda)^{1-d} \right) \left| B_{1/2} \setminus K_{2^k}(u) \right| ;$$

the induction hypothesis yields

$$\left| B_{1/2} \setminus K_{2^{k+1}}(u) \right| \leq \left(1 - 2^{-2} (2d\Lambda/\lambda)^{1-d} \right)^{k+1} \left| B_{1/2} \right| .$$

Step 4. We learned in Step 1 that

$$\left\{ u > 2^n \right\} \cap B_{1/2} \subset B_{1/2} \setminus K_{2^n} .$$

It follows that

$$\left| \left\{ u > 2^n \right\} \cap B_{1/2} \right| \leq \left| B_{1/2} \setminus K_{2^n} \right| \leq (1-\sigma)^n \left| B_{1/2} \right| ,$$

for every $n \in \mathbb{N}$, where $\sigma := 2^{-2} (2d\Lambda/\lambda)^{1-d}$.

Now, let $t > 0$. There exists $n \in \mathbb{N}$ such that $2^n < t < 2^{n+1}$. Choose $\varepsilon > 0$ such that $1 - \sigma = 2^{-\varepsilon}$. Hence,

$$\left| \{ u > t \} \cap B_{1/2} \right| \leq \frac{(1-\sigma)^{n+1}}{1-\sigma} \left| B_{1/2} \right| \leq \frac{t^{-\varepsilon}}{1-\sigma} \left| B_{1/2} \right| .$$

The result follows by choosing the (universal) constant $C(d, \lambda, \Lambda)$. □

To complete the proof of Theorem 1.29 it remains for us to derive a C^α-regularity result from the L^ε-estimate.

Proof of Theorem 1.29   Once Proposition 1.39 is available, we resort to standard arguments to conclude that

$$\operatorname{osc}_{B_r} u \le \gamma \operatorname{osc}_{B_{4r}} u,$$

for some universal $\gamma \in (0, 1)$; see, for instance, Gilbarg and Trudinger (2001, Chapter 8.9). An application of Lemma 1.30 concludes the proof.   □

1.3 Lin's Integral Estimates

The Krylov–Safonov theory provides a modulus of continuity for the viscosity solutions to inequalities of the form $\mathcal{P}^+_{\lambda, \Lambda} \ge 0$. However, those viscosity inequalities also lead to integral estimates. We study this class of results in what follows.

Lin (1986) examines $C^{1,1}$-solutions to the Dirichlet problem

$$\operatorname{Tr}\left(A(x)D^2 u\right) = -f \quad \text{in} \quad B_1, \tag{1.48}$$

satisfying $u \equiv 0$ on ∂B_1. The matrix $A\colon B_1 \to S(d)$ is measurable and (λ, Λ)-elliptic, and $f \in L^d(B_1)$.

Let $u \in C^{1,1}(B_1)$ be a solution to (1.48). For $t > 0$ define $E_t \subset B_1$ as

$$E_t := \left\{ x \in B_1 \mid \left| D^2 u(x) \right| > t \right\}.$$

The main result in Lin (1986) concerns a decay rate for the measure of E_t, with respect to $t > 0$. It ensures the existence of constants $C = C(\lambda, \Lambda, d)$ and $\varepsilon = \varepsilon(d, \lambda, \Lambda)$ such that

$$|E_t| \le C t^{-\varepsilon}.$$

Were $1 \le \varepsilon < \infty$, the standard theory of Lebesgue spaces would imply $D^2 u \in L^\varepsilon(B_1)$. This is not the case, since the explicit dependence of ε on the dimension and the ellipticity constants is unknown and ε could be very small (see Remark 1.41). In spite of this observation, Lin's integral estimate has been referred to as the $W^{2, \varepsilon}$-*estimates*.

Caffarelli and Cabré (1995) extend Lin's estimate to the fully nonlinear, inhomogeneous, setting, in the context of C-viscosity solutions. In what follows we detail Lin's integral estimate for viscosity solutions to

$$\mathcal{P}^+_{\lambda, \Lambda}\left(D^2 u\right) \ge 0 \quad \text{in} \quad \Omega. \tag{1.49}$$

Our argument follows closely the one developed by Armstrong et al. (2012). We proceed with a definition. For $u \in C(B_1)$, define

$$\underline{\Theta}(x) := \inf \left\{ A \geq 0 \mid \exists p \in \mathbb{R}^d; \, u(y) \geq u(x) + p \cdot (x - y) - \frac{A}{2}|x - y|^2, \, \forall y \in B_1 \right\}.$$

Given $x \in B_1$, the function $\underline{\Theta}$ accounts for the largest possible opening of a paraboloid touching the graph of u from below. We notice that $\underline{\Theta}$ is positively homogeneous of degree 1 with respect to u.

Theorem 1.40 (Lin's integral estimate)   *Let $u \in C(B_1)$ be a viscosity solution to (1.49). There exist constants $C = C(\lambda, \Lambda, d)$, $t_0 = t_0(\lambda, \Lambda, d)$, and $\varepsilon = \varepsilon(\lambda, \Lambda, d)$ such that*

$$\left| \{ x \in B_{1/2} \mid \underline{\Theta}(x) > t \} \right| \leq Ct^{-\varepsilon},$$

for every $t > t_0$.

Remark 1.41   Armstrong et al. (2012) find that $\varepsilon \to 0$ as $\Lambda/\lambda \to \infty$. The explicit expression

$$\varepsilon = \frac{1}{2^2} \left(\frac{\lambda}{2d\Lambda} \right)^{d-1}$$

appeared in Mooney (2019); the author notices it is optimal in the planar setting.

To establish Theorem 1.40 we combine two ingredients. Namely, a measure-theoretical lemma and a consequence of Calderón–Zygmund cube decomposition. We proceed by recalling the existence of a barrier function.

Lemma 1.42 (Barrier function)   *Let $0 < \lambda \leq \Lambda$ be fixed. There exist $\varphi, \xi \in C^\infty(\mathbb{R}^d)$ and constants $C > 0$ and $M > 1$ such that*

$$\mathcal{P}^-_{\lambda, \Lambda}(D^2\varphi) \geq -C\xi \quad in \quad \mathbb{R}^d.$$

In addition, $\varphi \geq -M$ in $\mathbb{R}^d$, $\varphi \leq -1$ in Q_3 and $\varphi \geq 0$ in $\mathbb{R}^d \setminus B_{6\sqrt{d}}$. Finally, $0 \leq \xi \leq 1$, with $\xi \equiv 0$ on $\mathbb{R}^d \setminus Q_1$.

For a proof of this result we refer the reader to Caffarelli and Cabré (1995, Lemma 4.1). We also recall a consequence of Calderón–Zygmund cube decomposition, as stated in Armstrong et al. (2012, Proposition 2.3).

Lemma 1.43   *Let $D \subset E \subset Q_1$ be measurable sets. Suppose $\delta \in (0,1)$ is such that $|D| \leq \delta|Q_1|$. Suppose further that, for $x \in \mathbb{R}^d$ and $r > 0$ satisfying*

$$Q_{x,3r} \subset Q_1 \quad and \quad \left| D \cap Q_{x,r} \right| \geq \delta|Q_r|,$$

we have $Q_{x,3r} \subset E$. Then $|D| \leq \delta|E|$.

We refer the reader to Caffarelli and Cabré (1995, Lemma 4.2) for a proof of Lemma 1.43. We proceed with a measure theoretical proposition. From now on, we suppose that $B_{6\sqrt{d}} \subset \Omega$.

Proposition 1.44 *Let $u \in C(\Omega)$ be a viscosity solution to (1.49). Suppose there exists $t > 0$ satisfying*

$$\{\underline{\Theta}(u, \Omega) \le t\} \cap Q_3 \ne \emptyset. \tag{1.50}$$

Then there exist constants $M > 1$ and $\sigma > 0$ such that

$$\left|\{\underline{\Theta}(u, \Omega) \le Mt\} \cap Q_1\right| \ge \sigma.$$

Proof We split the proof into three steps.

Step 1. Without loss of generality, we suppose $t = 1$. In fact, this is possible because the operator $\mathcal{P}^+_{\lambda, \Lambda}$ and the function $\underline{\Theta}(u, \cdot)(\cdot)$ are positively homogeneous of degree 1 with respect to u. Notice that adding an affine function to u does not affect either (1.49) or the information on Θ. Because of (1.50) we know there exists $x_0 \in Q_3$ such that a polynomial of opening 1 touches u from below at x_0. Let $P(x)$ given by

$$P(x) := \frac{1}{2}\left(36d - |x|^2\right)$$

satisfy

$$\inf_{x \in \Omega}(u(x) - P(x)) = u(x_0) - P(x_0) = 0.$$

It is clear that

$$u \ge P \ge 0$$

in $B_{6\sqrt{d}}$. Also,

$$u(x_0) = P(x_0) \le \sup_{x \in Q_3} P(x) = 18d.$$

Step 2. Let φ be the barrier function whose existence is ensured by Lemma 1.42. Define $w := u + A\varphi$, where $A > 0$ is a constant to be determined further. Hence,

$$\mathcal{P}^+(D^2 w) = \mathcal{P}^+(D^2 u + A D^2 \varphi) \ge \mathcal{P}^+(D^2 u) + A\mathcal{P}^-(D^2\varphi) \ge -AC\xi,$$

in $B_{6\sqrt{d}}$, with $w \ge 0$ on $\partial B_{6\sqrt{d}}$. Because $x_0 \in Q_3$, we have

$$\sup_{x \in B_{6\sqrt{d}}} w^-(x) \ge -(u(x_0) + A\varphi(x_0)) \ge -18d + A.$$

An application of the ABP maximum principle ensures that

$$A - 18d \leq CA|\{x \in Q_1 \mid \Gamma_w(x) = w(x)\}|^{\frac{1}{d}},$$

where Γ_w is the convex envelope of $-w^-\chi_{6\sqrt{d}}$ in $B_{12\sqrt{d}}$. By setting $A := 19d$ one recovers

$$\sigma \leq |\{x \in Q_1 \mid \Gamma_w(x) = w(x)\}|,$$

where the universal constant σ is given by

$$\sigma := \left(\frac{1}{19C}\right)^d.$$

To complete the proof it suffices to check there exists $M > 1$ such that

$$|\{x \in Q_1 \mid \Gamma_w(x) = w(x)\}| \subset |\{x \in Q_1 \mid \Theta(u,\Omega)(x) \leq M\}|;$$

this is the subject of the next step.

Step 3. Let $x_1 \in \{\Gamma_w = w\}$. We prove $x_1 \in \{\Theta(u,\Omega) \leq M\}$. Because $\Gamma_w \leq -w^-\chi_{B_6\sqrt{d}}$, it follows that $\Gamma_w < 0$ in $B_{12\sqrt{d}}$. Therefore we can find an affine function $\ell(x)$, with $\ell \leq 0$ in $B_{12\sqrt{d}}$, satisfying

$$\ell(x) - A\varphi(x) \leq \Gamma_w(x) - A\varphi(x) \leq u(x)$$

in $B_{6\sqrt{d}}$, with equality at x_1.

We claim there exist a positive constant $C = C(d, A, K)$ such that

$$|D\ell| \leq C$$

in $B_{12\sqrt{d}}$. It follows from $\ell \leq 0$ in that ball, combined with

$$\ell(x_1) = u(x_1) + A\varphi(x_1) \geq -KA.$$

Because $|D^2\varphi| < C$ for some universal constant $C > 0$, there exists a polynomial P^* of universal opening $M > 0$ satisfying

$$P^* \leq u$$

in $B_{6\sqrt{d}}$, with $P^*(x_1) = u(x_1)$. We notice that $P^* \leq P$ on $\partial B_{6\sqrt{d}}$; hence by taking $M \gg 1$ large enough, we ensure that $P^* \leq P$ in $\mathbb{R}^d \setminus B_{6\sqrt{d}}$. By gathering the former inequalities, we learn that $P^* \leq u$ in Ω, with equality at x_1 and the proof is complete.   □

In Proposition 1.44, the conclusion could be written as

$$|\{\Theta(u,\Omega) \leq Mt\} \cap Q_1| \geq \sigma|Q_1|.$$

Also, we used the cubes Q_3 and Q_1. The same result is available if we consider instead $Q_{x,3r}$ and $Q_{x,r}$, respectively, for $x \in \Omega$ and $r > 0$ satisfying $Q_{x,3r} \subset \Omega$. In this case, the conclusion would read

$$\left| \{ \underline{\Theta}(u, \Omega) \leq Mt \} \cap Q_{x,r} \right| \geq \sigma \left| Q_{x,r} \right|.$$

We close this section with the proof of Theorem 1.40.

Proof of Theorem 1.40 We combine Proposition 1.44 with Lemma 1.43. Because u is bounded in B_1, there exists $C > 0$ and $x \in Q_{1/2\sqrt{d}}$ such that

$$\underline{\Theta}(u, B_1)(x) \leq C \sup_{x \in B_1} |u(x)|.$$

Hence, there exists $t_0 > 0$ such that for every $t > t_0$

$$\{ \underline{\Theta}(u, B_1)(x) \leq t \} \cap Q_{1/2\sqrt{d}} \neq \emptyset. \tag{1.51}$$

Step 1. Fix $t > t_0$ and define

$$D := \{ \underline{\Theta}(u, B_1) > Mt \} \cap Q_{1/6\sqrt{d}}$$

and

$$E := \{ \underline{\Theta}(u, B_1) > t \} \cap Q_{1/6\sqrt{d}}.$$

Because of (1.51), Proposition 1.44 yields

$$\left| \{ \underline{\Theta}(u, B_1) > Mt \} \cap Q_{1/6\sqrt{d}} \right| \leq (1 - \sigma) \left| Q_{1/6\sqrt{d}} \right|;$$

by setting $\delta := (1 - \sigma)$ in Lemma 1.43, we conclude $|D| \leq \delta \left| Q_{1/6\sqrt{d}} \right|$. It remains for us to check that D and E satisfy the second condition in Lemma 1.43. This is done in the next step.

Step 2. The contrapositive of Proposition 1.44 ensures that, for every $t > 0$ satisfying

$$\left| \{ \underline{\Theta}(u, B_1) > Mt \} \cap Q_{x,r} \right| > (1 - \sigma) \left| Q_{x,r} \right|,$$

we have

$$\underline{\Theta}(u, B_1)(x) > t,$$

for every $x \in Q_{x,3r}$. That is, $Q_{x,3r} \subset E$. It follows from Lemma 1.43 that $|D| \leq (1 - \sigma)|E|$. That is,

$$\left| \{ \underline{\Theta}(u, B_1) > Mt \} \cap Q_{1/6\sqrt{d}} \right| \leq (1 - \sigma) \left| \{ \underline{\Theta}(u, B_1) > t \} \cap Q_{1/6\sqrt{d}} \right|.$$

An scaling argument and the homogeneity of $\Theta(u, \cdot)(\cdot)$ build upon the former inequality to produce

$$\left|\left\{\underline{\Theta}(u, B_1) > M^k t\right\} \cap Q_{1/6\sqrt{d}}\right| \leq (1 - \sigma)\left|\left\{\underline{\Theta}(u, B_1) > M^{k-1} t\right\} \cap Q_{1/6\sqrt{d}}\right|, \tag{1.52}$$

for every $k \in \mathbb{N}$.

Step 3. Given $\tau \geq t$, there exists $k \in \mathbb{N}$ such that $M^{k-1} t \leq \tau < M^k t$. Fix $\varepsilon > 0$ to be determined further. We have

$$\begin{aligned}
\left|\left\{\underline{\Theta}(u, B_1) > \tau\right\} \cap Q_{x,r}\right| &\leq \left|\left\{\underline{\Theta}(u, B_1) > M^{k-1} t\right\} \cap Q_{x,r}\right| \\
&\leq (1 - \sigma)^{k-1} \tau^\varepsilon \tau^{-\varepsilon} \\
&\leq (1 - \sigma)^{k-1} \left(M^k t\right)^\varepsilon \tau^{-\varepsilon} \\
&\leq C\tau^{-\varepsilon},
\end{aligned}$$

where $C > 0$ is a universal constant, provided $0 < \varepsilon \ll 1$ is chosen such that

$$\left((1 - \sigma)M^\varepsilon\right)^k \leq 1.$$

At this point, a covering argument completes the proof. $\qquad\qquad\square$

1.4 Gradient Hölder Continuity

In this section we combine structural properties of fully nonlinear operators with the Krylov–Safonov theory to produce a Hölder regularity result for the solutions to

$$F\left(D^2 u\right) = 0 \quad \text{in} \quad \Omega. \tag{1.53}$$

We proceed by recalling an auxiliary result. It connects an estimate on the second-order increment quotient of a function $v \in C(B_1)$ with its smoothness in Hölder spaces.

Proposition 1.45 *Let $u \in C(\Omega) \cap L^\infty(\Omega)$. Suppose there exist $C > 0$ and $\alpha \in (0, 1)$ such that*

$$\sup_{x \in B_r(x_0)} |u(x_0 + x) + u(x_0 - x) - 2u(x_0)| \leq Cr^{1+\alpha},$$

for every $0 < r < 1$ and every $x_0 \in \Omega$, with $B_r(x_0) \subset \Omega$. Then $u \in C_{loc}^{1,\alpha}(\Omega)$. In addition, for every $\Omega' \Subset \Omega$ there exists $C_0 > 0$ for which

$$\|u\|_{C^{1,\alpha}(\Omega')} \leq C_0 \left(\|u\|_{L^\infty(\Omega)} + C\right).$$

For a proof of Proposition 1.45 we refer the reader to Stein (1970, Propositions 8 and 9, Chapter 5). For its use in a context similar to the present one, we refer to Trudinger (1988, 1989).

Before stating the main theorem in this section, we discuss a simplifying assumption. Let F be a (λ, Λ)-elliptic operator and define $F_0 = F_0(M)$ as

$$F_0(M) := F(M) - F(0);$$

it is clear that F_0 is a (λ, Λ)-elliptic operator. As a consequence, in what follows, we suppose $F(0) = 0$. The main theorem in this section is the following.

Theorem 1.46 (Gradient Hölder continuity) *Let $u \in C(\Omega)$ be a viscosity solution to (1.53). Suppose F is a (λ, Λ)-elliptic operator satisfying $F(0) = 0$. Then $u \in C_{loc}^{1,\alpha}(\Omega)$, for some $\alpha \in (0,1)$ depending only on d, λ, Λ and $\mathrm{diam}(\Omega)$. Moreover, for every $\Omega' \Subset \Omega$, we have*

$$\|u\|_{C^{1,\alpha}(\Omega')} \leq C \|u\|_{L^\infty(\Omega)},$$

where $C = C(d, \lambda, \Lambda, \mathrm{diam}(\Omega), \mathrm{dist}(\Omega', \partial))$ is a nonnegative constant.

Proof The proof relates (1.53) with the extremal operator $\mathcal{P}^+$. It applies the Krylov–Safonov theory to an auxiliary function, yet to be introduced. This fact yields a bound on the second-order increment quotient associated with u, and the result follows from a straightforward application of Proposition 1.45.

Step 1. Fix a direction $\theta \in \mathbb{S}^{d-1}$ and consider $x + h\theta$, for $0 < h \ll 1$ and $x \in \Omega$. Because u solves (1.53), the function $u(x + h\theta)$ solves $F(D^2 u(x + h\theta)) = 0$ in Ω_{2h}, where

$$\Omega_{2h} := \{x \in \Omega \mid \mathrm{dist}(x, \partial\Omega) > 2h\}.$$

As a consequence,

$$\frac{1}{h} F\left(\frac{h}{h} D^2 u(x + h\theta)\right) - \frac{1}{h} F\left(\frac{h}{h} D^2 u(x)\right) = 0.$$

Notice that $G_h : S(d) \to \mathbb{R}$ defined as

$$G_h(M) := \frac{1}{h} F(hM)$$

has the same ellipticity constants as F. As a conclusion,

$$0 = G_h\left(\frac{D^2 u(x + h\theta)}{h}\right) - G_h\left(\frac{D^2 u(x)}{h}\right) \leq \mathcal{P}^+\left(\frac{D^2 u(x + h\theta) - D^2 u(x)}{h}\right)$$

in Ω_{2h}. As a consequence of Theorem 1.29, we infer that

$$v_h(x) := \frac{u(x + h\theta) - u(x)}{h}$$

is α-Hölder continuous for some $\alpha \in (0, 1)$, not depending either on h or on θ, with universal estimates.

Step 2. The modulus of continuity and the estimates available for v_h imply that

$$\left| \frac{u(x + h\theta) - u(x)}{h} - \frac{u(y + h\theta) - u(y)}{h} \right| \leq C|x - y|^{1+\alpha},$$

for every $x, y \in \Omega_{2h}$. By choosing $y := x - h\theta$, the former inequality becomes

$$|u(x + h\theta) + u(x - h\theta) - 2u(x)| \leq Ch^{1+\alpha}.$$

As a conclusion, we obtain

$$\sup_{x \in B_r(x_0)} |u(x_0 + x) + u(x_0 - x) - 2u(x_0)| \leq Cr^{1+\alpha},$$

for every $0 < r \ll 1$ and $x_0 \in \Omega$ such that $B_r(x_0) \subset \Omega$. An application of Proposition 1.45 finishes the proof. □

Remark 1.47 (On the condition $F(0) = 0$)   If we drop the assumption $F(0) = 0$ the statement of Theorem 1.46 remains the same, except for the estimate available for the solutions. In the general case, it would read

$$\|u\|_{C^{1,\alpha}(\Omega')} \leq C \left(\|u\|_{L^\infty(\Omega)} + |F(0)| \right). \qquad (1.54)$$

This observation follows from Theorem 1.56.

1.5 Evans–Krylov Theory

We have learned three fundamental facts on the regularity theory associated with (λ, Λ)-elliptic operators. First, if $u \in C(\Omega)$ is a viscosity solution to

$$\mathcal{P}^+_{\lambda, \Lambda}(D^2u) \geq 0 \geq \mathcal{P}^-_{\lambda, \Lambda}(D^2u) \quad \text{in} \quad \Omega \qquad (1.55)$$

the Krylov–Safonov theory implies $u \in C^\alpha_{\text{loc}}(\Omega)$, with the appropriate estimates, for some $\alpha = \alpha(d, \lambda, \Lambda) \in (0, 1)$. Also, Lin's integral estimates, as stated in Theorem 1.40, are available for the solutions to (1.55). In addition, let $u \in C(\Omega)$ be an L^p-viscosity solution to

$$F(x, D^2u) = 0 \quad \text{in} \quad \Omega,$$

where F is a (λ, Λ)-elliptic operator. In this case, we infer that solutions are differentiable, with Hölder-continuous gradient; i.e., $u \in C^{1,\alpha}_{\mathrm{loc}}(\Omega)$, for some $\alpha \in (0, 1)$, depending only on the dimension and the ellipticity constants.

In this section we impose a further condition on the operator F and examine its effect on the regularity of the solutions to

$$F(D^2 u) = 0 \quad \text{in} \quad \Omega. \tag{1.56}$$

In what follows we suppose the (λ, Λ)-elliptic F to be convex with respect to the Hessian of the solutions and prove that viscosity solutions are locally of class $C^{2,\alpha}$.

Known as the Evans–Krylov theory, the argument towards $C^{2,\alpha}$-regularity for the solutions to equations in the nondivergence form was first established by Evans (1982a) and Krylov (1982), independently, in the early 1980s. Here we detail the proof presented by Caffarelli and Cabré (1995, Chapter 6). A shorter proof was recently presented by Caffarelli and Silvestre (2010a); it is inspired by developments due to those authors in the nonlocal setting. We continue by stating the main result in this section.

Theorem 1.48 (Evans–Krylov Regularity Theorem) *Let $u \in C(\Omega)$ be an L^p-viscosity solution to (1.56). Suppose the operator $F : S(d) \to \mathbb{R}$ is (λ, Λ)-elliptic and concave; suppose further $F(0) = 0$. Then $u \in C^{2,\alpha}_{loc}(\Omega)$, for some universal $\alpha \in (0, 1)$. In addition, for every $\Omega' \Subset \Omega$ there exists $C > 0$ such that*

$$\|u\|_{C^{2,\alpha}(\Omega')} \le C \|u\|_{L^\infty(\Omega)},$$

where $C = C(d, \lambda, \Lambda, \mathrm{diam}(\Omega), \mathrm{dist}(\Omega', \partial\Omega))$.

The proof of Theorem 1.48 relies on three main facts. First we prove that solutions to $F = 0$ are $C^{1,1}$-regular. Then we establish oscillation control for a quantity depending on the Hessian of the solutions. Once this information is available, a scaling argument builds upon Lemma 1.30 and leads to the result.

For $\Omega_1 \subset \Omega$, denote with $D^2 u(\Omega_1)$ the set

$$D^2 u(\Omega_1) := \left\{ M \in S(d) \mid M = D^2 u(x) \text{ for some } x \in \Omega_1 \right\}.$$

Because u solves (1.56) it is twice-differentiable almost everywhere; as a consequence, $D^2 u(x)$ is well defined for almost every $x \in \Omega$. Thus we may regard $D^2 u(\Omega)$ as the essential image of Ω by the Hessian of the solutions. We are interested in the oscillation of the essential diameter of sets $D^2 u(\Omega_1)$.

We observe that any $\Omega' \Subset \Omega$ can be covered by a finite number m of balls $B_{r/2}(x_1), \ldots, B_{r/2}(x_m)$, with $x_i \in \Omega'$ for $i = 1, \ldots, m$, and $r \in (0, 1)$ fixed, satisfying $\overline{B_r(x_i)} \subset \Omega$. For that reason, we examine (1.56) with $\Omega \equiv B_1$

and study the interior regularity of the solutions in $B_{1/2}$. We continue with an auxiliary lemma.

Lemma 1.49 (On the difference of super and subsolutions) *Let $u \in C(B_1)$ be an L^d-viscosity supersolution to (1.56) and $v \in C(B_1)$ be an L^d-viscosity subsolution to (1.56). Suppose F is a (λ, Λ)-elliptic operator. Then*

$$\mathcal{P}^+(D^2u - D^2v) \geq 0 \quad in \quad B_1.$$

For the proof of Lemma 1.49 we refer the reader to Caffarelli and Cabré (1995, Theorem 5.3). Next we prove that solutions to (1.56) are of class $C^{1,1}$, provided F is (λ, Λ)-elliptic and concave.

Proposition 1.50 (Estimates in $C^{1,1}$-spaces) *Let $u \in C(B_1)$ be an L^p-viscosity solution to (1.56). Suppose the operator $F \colon S(d) \to \mathbb{R}$ is (λ, Λ)-elliptic and concave; suppose further $F(0) = 0$. Then $u \in C^{1,1}_{loc}(B_1)$ and there exists $C > 0$ such that*

$$\|D^2u\|_{L^\infty(B_{1/2})} \leq C\|u\|_{L^\infty(B_1)}.$$

Moreover, $C = C(d, \lambda, \Lambda)$.

Proof For ease of presentation, we split the proof into 3 steps.

Step 1. We start by noticing that u is subharmonic. Indeed, because F is concave and fixes the origin, we have

$$\Delta\varphi(x) \geq F(D^2\varphi(x))$$

in B_1, for every $\varphi \in C^2(B_1)$. As a consequence, $\Delta u \geq 0$ in the viscosity sense in the unit ball. For every $x_0 \in B_{1/2}$ and $0 < h \ll 1$, let $w \in C(B_h(x_0))$ be a viscosity solution to $\Delta w = 0$ in $B_h(x_0)$, agreeing with u on $\partial B_h(x_0)$. Since $\Delta u \geq \Delta w$ in $B_h(x_0)$ and $u = w$ on $\partial B_h(x_0)$, we infer that

$$u(x_0) \leq w(x_0) \leq \fint_{\partial B_h(x_0)} w\, dS = \fint_{\partial B_h(x_0)} u\, dS.$$

Next, define the nonnegative function u_h^* as

$$u_h^*(x) := \frac{1}{h^2}\fint_{\partial B_h(x)} u(y) - u(x)dS(y);$$

we study the function u_h^* and produce an estimate for D^2u in $L^2(B_{9/10})$. Let $\varphi \in C^2(B_1)$ and compute its Taylor polynomial of order two, centered at the origin. That is,

$$\varphi(x) - \varphi(0) = D\varphi(0) \cdot x + \frac{1}{2}x^T D^2\varphi(0)x.$$

By integrating the former equality over ∂B_h one gets

$$\int_{\partial B_h} \varphi(x) - \varphi(0) dS(x) = \frac{h^2}{2d} |\partial B_h| \, \Delta\varphi(0) + o(h).$$

As a consequence, we notice that $\varphi_h^*(x)$ converges locally uniformly to the quantity $(2d)^{-1}\Delta\varphi(x)$, and

$$\left\| \varphi_h^* \right\|_{L^\infty(B_{1/2})} \le C \left\| D^2\varphi \right\|_{L^\infty(B_1)},$$

where $C > 0$ depends only on the dimension d. The connection between u_h^* and Δu is central to our argument. We proceed with elementary observations; notice that u_h^* is locally bounded in the L^1-norm. Indeed, let $\phi \in C^\infty(B_1)$ be nonnegative and compactly supported in $B_{8/9}$, with $\phi \equiv 1$ in $B_{7/8}$. Hence

$$\int_{B_{7/8}} \left| u_h^*(x) \right| dx = \int_{B_{7/8}} u_h^*(x) dx \le \int_{B_{8/9}} u_h^*(x)\psi(x) dx = \int_{B_{8/9}} u(x)\psi_h^*(x) dx,$$

where the second equality follows from a change-of-variables argument. In conclusion,

$$\left\| u_h^* \right\|_{L^1(B_{7/8})} \le C \left\| D^2\phi \right\|_{L^\infty(B_1)} \le C.$$

Step 2. In the sequel we verify that u_h^* solves

$$\mathcal{P}^- \left(D^2 u_h^* \right) \le 0 \quad \text{in} \quad B_{7/8}. \tag{1.57}$$

We start by noticing that

$$\fint_{\partial B_h(x)} u(y) dy = \fint_{\partial B_h} u(x+y) dy;$$

for $k \in \mathbb{N}$, choose $y_1, \ldots, y_k$ points in ∂B_h and set

$$u_i(x) := u(x + y_i).$$

The convex combination

$$v_k^*(x) := \frac{1}{k} u_1(x) + \cdots + \frac{1}{k} u_k(x)$$

satisfies

$$F\left(D^2 v_k^* \right) \ge \frac{1}{k} \sum_{i=1}^k F\left(D^2 u_i \right) = 0 \quad \text{in} \quad B_{7/8},$$

since F is concave. In addition,

$$v_k^*(\cdot) \longrightarrow \fint_{\partial B_h(\cdot)} u(y) dy,$$

locally uniformly, as $k \to \infty$. The stability of viscosity supersolutions ensures that

$$\fint_{\partial B_h(x)} u(y) dy$$

is a viscosity supersolution to $F = 0$ in $B_{7/8}$. Because u is a subsolution to $F = 0$, Lemma 1.49 implies (1.57). The next step is to bound the second-order increment of u.

Step 3. Because u_h^* solves (1.57) and we have a uniform bound for u_h^* in $L^1(B_{8/9})$, we infer that

$$\|u_h^*\|_{L^\infty(B_{7/8})} \leq C,$$

for a universal constant $C > 0$. Let $\psi \in C^\infty(B_1)$ be compactly supported in $B_{7/8}$. The former estimate yields

$$\left| \int_{B_{7/8}} u(x) \Delta \psi(x) dx \right| \leq 2d \lim_{h \to 0} \int_{B_{7/8}} |u \psi_h^*| dx \leq C \|u_h^*\|_{L^\infty(B_{7/8})} \|\psi\|_{L^1(B_1)} \leq C;$$

hence,

$$\|\Delta u\|_{L^\infty(B_{7/8})} \leq C.$$

As a consequence of the L^q-regularity estimates for the Poisson equation we infer that

$$\left\| D^2 u \right\|_{L^q(B_{7/8})} \leq C,$$

for every $1 < q < \infty$. Denote by $\Delta_{h,\theta} u(x)$ the second-order h-increment of u in the direction $\theta \in \mathbb{S}^{d-1}$ given by

$$\Delta_{h,\theta} u(x) := \frac{u(x + h\theta) + u(x - h\theta) - 2u(x)}{h^2}.$$

The L^q-bound on the Hessian of u implies that

$$\left\| \Delta_{h,\theta} u \right\|_{L^q(B_{6/7})} \leq C, \tag{1.58}$$

for some universal constant $C > 0$. As before, we notice that

$$\mathcal{P}^- \left(D^2 \Delta_{h,\theta} u \right) \leq 0 \quad \text{in} \quad B_{7/8}.$$

Applying once again the L^∞-bounds available for subsolutions of $\mathcal{P}^- = 0$ and using (1.58) one obtains

$$\sup_{x \in B_{6/7}} \Delta_{h,\theta} u(x) \leq C, \tag{1.59}$$

uniformly in $0 < h \ll 1$ and $\theta \in \mathbb{S}^{d-1}$. It follows that u is semiconcave and, as a consequence of Proposition 1.25, it is twice-differentiable almost everywhere. From (1.59) we infer

$$\partial_{\theta\theta} u(x) \leq C, \tag{1.60}$$

uniformly in $x \in B_{6/7}$ and $\theta \in \mathbb{S}^{d-1}$. Because u is twice-differentiable almost everywhere, we have $F(D^2 u(x)) = 0$ almost everywhere in $B_{6/7}$. Applying Corollary 1.9 with $M := D^2 u(x)$ and $N := 0$ we get

$$\left\| D^2 u(x) \right\| \leq C \sup_{\theta \in \mathbb{S}^{d-1}} \left(\partial_{\theta\theta} u(x) \right)^+,$$

for almost every $x \in B_{6/7}$. By combining the former inequality with (1.60) we obtain

$$\left\| D^2 u \right\|_{L^\infty(B_{6/7})} \leq C$$

and conclude the proof. $\qquad\square$

In what follows we prove an intermediate proposition. It ensures that it is possible to refine the covering of the essential image of $D^2 u$, when moving from $B_{6/7}$ to $B_{3/7}$.

Proposition 1.51 *Let $u \in C(B_1)$ be an L^p-viscosity solution to (1.56) and $F: S(d) \to \mathbb{R}$ be a (λ, Λ)-elliptic, concave, operator. Suppose that*

$$1 < \mathrm{ess\,diam} D^2 u(B_{6/7}) \leq 2;$$

suppose further that there exist balls $B^1, \dots, B^N$ of radius $0 < \varepsilon < \varepsilon_0$ in $S(d)$ such that

$$D^2 u(x) \subset B^1 \cup \dots \cup B^N,$$

for almost every $x \in B_{6/7}$. Then, there exist $N - 1$ balls among the collection $(B^i)_{i=1}^N$ such that

$$D^2 u(x) \subset B^1 \cup \dots \cup B^{N-1},$$

for almost every $x \in B_{3/7}$. The constant $\varepsilon_0 > 0$ is universal.

Proof The proof consists of two building blocks. First we control the difference of the directional Hessian of u at two distinct points by a universal constant. Then we use the $C^{1,1}$-estimate for the solutions to prove that one particular intersection has measure zero.

Step 1. We learned in the proof of Proposition 1.50 that u is twice-differentiable almost everywhere in $B_{6/7}$. Denote with $A \subset B_{6/7}$ the set of points

$$A := \{x \in B_{6/7} \mid D^2 u(x) \text{ exists at } x\}.$$

It is clear that $|A| = |B_{6/7}|$.

For each ball B^i, take $x_i \in A$ such that $B^i \subset B_{2\varepsilon}(D^2 u(x_i))$. For ease of presentation, set $M_i := D^2 u(x_i)$, for $i = 1, \ldots, N$. Define the constant $c_0 := \lambda/(\Lambda - \lambda)$ and take ε_0 such that

$$2\varepsilon \leq 2\varepsilon_0 \leq \frac{c_0}{16}.$$

Then

$$D^2 u(x) \subset B_{c_0/16}(M_1) \cup \cdots \cup B_{c_0/16}(M_N),$$

for almost every $x \in B_{6/7}$. Because the essential diameter of $B_{6/7}$ is bounded from above by 2, there exists a ball $B \subset S(d)$ of radius 2 such that $M_i \subset B$ for every $i = 1, \ldots, N$.

Let $N' \leq N$ denote the number of points M_i such that

$$\|M_i - M_j\| \geq \frac{c_0}{16}. \tag{1.61}$$

That is, reorganize the family $(M_i)_{i=1}^{N}$ so that any two matrices in $(M_1, \ldots, M_{N'})$ meet (1.61) for every $i, j \in \{1, \ldots, N'\}$. We claim that

$$D^2 u(x) \subset B_{c_0/8}(M_1) \cup \cdots \cup B_{c_0/8}(M_{N'})$$

for almost every $x \in B_{6/7}$. Indeed, set $M := D^2 u(x)$; if $M \in B_{c_0/16}(M_i)$ for some $i = 1, \ldots, N'$, it is clear that $M \in B_{c_0/8}(M_i)$. Suppose otherwise: $M \in B_{c_0/16}(M_i)$ for some $i = N' + 1, \ldots, N$. For every $j = 1, \ldots, N'$ we obtain

$$\left|M - M_j\right| \leq \left|M - M_i\right| + \left|M_i - M_j\right| \leq \frac{c_0}{8}$$

and the claim is verified.

As a consequence, for almost every $x \in B_{6/7}$ we have

$$x \in \left(D^2 u\right)^{-1}\left(B_{c_0/8}(M_1)\right) \cup \cdots \cup \left(D^2 u\right)^{-1}\left(B_{c_0/8}(M_{N'})\right).$$

Hence, there exists $i^* = 1, \ldots, N'$ such that

$$\left|\left(D^2 u\right)^{-1}(B_{c_0/8}(M_{i^*})) \cap B_{3/7}\right| \geq \eta,$$

for some $\eta > 0$ universal. For simplicity, set $i^* := 1$.

By further refining the choice of ε such that $2\varepsilon \leq 2\varepsilon_0 \leq 1/4$ we ensure the existence of M_2 such that $\|M_1 - M_2\| \geq 1/4$. In fact, if such M_2 were not available, it would be impossible to have

$$D^2 u(x) \subset B_{2\varepsilon}(M_1) \cup \cdots \cup B_{2\varepsilon}(M_N),$$

for almost every $x \in B_{6/7}$.

An application of Corollary 1.9 ensures that

$$\frac{c_0}{4} \leq \sup_{\theta \in \mathbb{S}^{d-1}} \theta^T (M_2 - M_1)\theta.$$

Hence, there exists $\theta \in \mathbb{S}^{d-1}$ for which

$$\frac{c_0}{4} + u_{\theta\theta}(x_1) \leq u_{\theta\theta}(x_2). \qquad (1.62)$$

Step 2. In what follows, (1.62) builds upon the $C^{1,1}$-estimate available for u to refine the covering $B^1, \ldots, B^N$. We start by noticing that

$$\Delta_{h,\theta} u(x) = \int_{-1}^{1} \partial_{\theta\theta} u(x + \tau h\theta)(1 - |\tau|)d\tau \leq \operatorname*{ess\,sup}_{x \in B_{6/7}} \partial_{\theta\theta} u(x) =: K.$$

Define the function w_h as

$$w_h(x) := K - \Delta_{h,\theta}(x),$$

and observe that $w_h \geq 0$ is continuous in $B_{5/6}$. In addition, $\mathcal{P}_{\lambda/n,\Lambda}^{-}(D^2 w_h) \geq 0$; hence, there exists a universal $p^* > 0$ such that

$$\left\| K - \Delta_{h,\theta} u \right\|_{L^{p^*}(B_{4/5})} \leq C\big(K - \Delta_{h,\theta} u(x)\big),$$

for every $x \in B_{4/5}$. By taking the limit inferior $h \to 0$ in the former inequality and applying Fatou's Lemma, we get

$$\left\| K - \partial_{\theta\theta} u \right\|_{L^{p^*}(B_{4/5})} \leq C\big(K - \partial_{\theta\theta} u(x)\big) \leq C, \qquad (1.63)$$

for every $x \in B_{4/5} \cap A$, where the second inequality follows from the $C^{1,1}$-estimate available for u. Now we produce a lower bound for the norm in (1.63). Arguing as before we have

$$\big| \{ K - \partial_{\theta\theta} \geq c_0/8 \} \cap B_{4/5} \big| \geq \eta;$$

hence,

$$\left\| K - \partial_{\theta\theta} u \right\|_{L^{p^*}(B_{4/5})} \geq \frac{c_0}{8} \eta^{1/p^*}.$$

Because $K - \partial_{\theta\theta} \geq 0$, those upper and lower bounds imply the existence of a universal constant $C_1 > 0$ such that

$$\operatorname*{ess\,inf}_{x \in B_{4/5}} K - \partial_{\theta\theta}(x) \geq C_1.$$

Because

$$D^2 u(x) \subset B_{2\varepsilon}(M_1) \cup \cdots \cup B_{2\varepsilon}(M_N),$$

for almost every $x \in B_1$, there exists $j^* \in \{1, \ldots, N\}$ such that

$$K - \partial_{\theta\theta} u(x_{j^*}) \leq 3\varepsilon.$$

We further refine the choice of ε_0 to ensure that $5\varepsilon \leq 5\varepsilon_0 \leq C_1$. Hence, $D^2 u(x) \notin B_{2\varepsilon}(M_{j^*})$ for almost every $x \in B_{4/5}$ and the proof is complete. □

Proposition 1.52 *Let $u \in C(\Omega)$ be an L^p-viscosity solution to (1.56). Suppose the operator F is (λ, Λ)-elliptic and concave. Suppose further*

$$\operatorname{ess diam}\big(D^2 u(B_1)\big) = 2.$$

There exists $\delta_0 \in (0, 1)$ such that

$$\operatorname{ess diam}\big(D^2 u(B_{\delta_0})\big) \leq 1.$$

Proof There exists a ball B, of diameter 2 in $S(d)$, such that $D^2 u(x) \in B$ for almost every $x \in B_1$. Therefore, we can find N balls $B^1, \ldots, B^N$, of radius $\varepsilon_0 > 0$ such that

$$D^2 u(x) \subset B^1 \cup \cdots \cup B^N$$

for almost every $x \in B_1$. It follows from Proposition 1.51 that

$$D^2 u(x) \subset B^1 \cup \cdots \cup B^{N-1}$$

for almost every $x \in B_{1/2}$. Suppose $\operatorname{ess diam} D^2 u(B_{1/2}) > 1$; we show that

$$D^2 u(x) \subset B^1 \cup \cdots \cup B^{N-2} \quad \text{for} \quad \text{a.e.-}x \in B_{1/4}. \qquad (1.64)$$

In fact, let $w(x)$ be defined as $w(x) := 4u(x/2)$; we have $F\big(D^2 w\big) = 0$ in B_1. If $\operatorname{ess diam} D^2 u(B_{1/2}) > 1$, we have

$$1 < \operatorname{ess diam} D^2 w(B_1) \leq 2.$$

Because of Proposition 1.51,

$$D^2 w(x) \subset B^1 \cup \cdots \cup B^{N-2},$$

for almost every $x \in B_{1/2}$, and (1.64) follows.

If $\operatorname{ess diam} D^2 u(B_{1/4}) > 1$, we repeat the process; by considering $w(x) := 16u(x/4)$, we find that

$$D^2 u(x) \subset B^1 \cup \cdots \cup B^{N-3},$$

for almost every $x \in B_{1/8}$. Because the number N of balls we begin with is finite, there exists $k \leq N$ such that $\operatorname{ess diam} D^2 u(B_{1/2^k}) \leq 1$. By taking

$$\delta_0 := \frac{1}{N} \leq \frac{1}{k},$$

the proof is complete. □

Next we put forward the proof of Theorem 1.48; it follows from the scaled counterpart of Proposition 1.52.

Proof of Theorem 1.48 Let $w \colon [0, 1] \to \mathbb{R}$ be defined as

$$w(r) := \operatorname{ess\,diam} D^2 u(B_r).$$

For the universal constant δ_0 in Proposition 1.52, a scaled version of this proposition yields

$$w(\delta_0 r) \le \frac{1}{2} w(r).$$

Then a straightforward application of Lemma 1.30 completes the proof.   □

1.6 Caffarelli's Regularity Theory: Approximation Methods

In this section we start discussing the use of approximation methods in the study of regularity theory. We consider a uniformly elliptic operator with variable coefficients governing an inhomogeneous problem. To be precise, we are interested in L^p-viscosity solutions to

$$F\left(x, D^2 u\right) = f \quad \text{in} \quad \Omega, \tag{1.65}$$

where F is a (λ, Λ)-elliptic operator and $f \in L^p(\Omega)$, $p > d$.

For every $x_0 \in \Omega$ fixed, the operator $F_{x_0}(M) := F(x_0, M)$ is (λ, Λ)-elliptic. It follows from Theorem 1.46 that L^p-viscosity solutions to

$$F_{x_0}\left(D^2 v\right) = 0 \quad \text{in} \quad \Omega \tag{1.66}$$

are $C^{1,\alpha}$-regular, and the appropriate estimates are available. The fundamental idea underlying the use of approximation methods here is to connect (1.65) with (1.66). Ideally, one is capable of transmitting information from the latter to the former. In this concrete example, the interesting information concerns the Hölder continuity of the gradient.

This analysis was introduced by Caffarelli (1989), where a regularity program in Hölder and Sobolev spaces is developed. These ideas have been detailed by Caffarelli (1988) in the linear setting. We also refer to Caffarelli and Cabré (1995).

To proceed we need to impose further conditions on the operator F and the source term f. From a heuristic viewpoint, these conditions encode the approximation regime relating both equations.

Definition 1.53 (Oscillation measure)   Let $F \colon \Omega \times S(d) \to \mathbb{R}$. For $x_0 \in \Omega$ define $\beta_{x_0} \colon \Omega \to \mathbb{R}$ as

$$\beta_{x_0}(x) := \sup_{M \in S(d)} \frac{\left| F(x_0, M) - F(x, M) \right|}{1 + \|M\|}.$$

The function in Definition 1.53 first appeared in Caffarelli (1989); see also Caffarelli and Cabré (1995). The first assumption under which we work concerns the smallness of the oscillation measure in L^d-spaces.

Assumption 1.54 (L^p-estimate for the oscillation)   We suppose $\beta_{x_0} \in L^p(\Omega)$ for every $x_0 \in \Omega$, for some $p > d$. In addition, there exists $0 < \theta \ll 1$, to be determined further, such that

$$\fint_{B_r(x_0)} |\beta_{x_0}(x)|^p \, dx \le \theta^p,$$

for every $x_0 \in \Omega$ and $0 < r \ll 1$ with $B_r(x_0) \subset \Omega$.

The parameter θ depends only on universal quantities; we determine it in the proof of Proposition 1.60, when setting the quantities in (1.77). We continue with a condition on the source term f.

Assumption 1.55 (Integrability of the source term)   We suppose $f \in L^p(\Omega)$, for some $p > d$. In addition, there exists $C > 0$ satisfying

$$\int_\Omega |f(x)|^p \, dx \le C.$$

By controlling the oscillation of the general operator with respect to its fixed-coefficients counterpart we devise an approximation method. It connects the equations governed by both operators allowing us to transmit information from the simpler one to the other.

Throughout this section – and in several other instances in this book – we suppose u is a *normalized* viscosity solution. It means that $\|u\|_{L^\infty(B_1)} \le 1$. We also require the L^p-norm of f to be sufficiently small. In the sequel we resort to a scaling argument to verify that such requirements do not represent any further restriction on the problem. On the contrary, they follow from the structure of (1.65) and Assumption 1.55. To see this is the case, define

$$v(x) := \frac{u(rx)}{K},$$

for constants $r, K > 0$ to be defined further. Notice that v solves

$$\overline{F}(x, D^2 v) = \overline{f}(x) \quad \text{in} \quad B_1,$$

where

$$\overline{F}(x, M) := \frac{r^2}{K} F\left(rx, \frac{K}{r^2}M\right)$$

and

$$\overline{f}(x) := \frac{r^2}{K} f(rx).$$

It is clear that $\overline{F}$ is a (λ, Λ)-elliptic operator. For arbitrary $\varepsilon \in (0, 1)$, set $r := \varepsilon$. Also, take

$$K := 1 + \|u\|_{L^\infty(B_1)} + \|f\|_{L^p(B_1)};$$

this choice is possible because viscosity solutions to (1.65) are universally bounded in $L^\infty(B_1)$ and Assumption 1.55. Clearly $\|v\|_{L^\infty(B_1)} \le 1$. Also,

$$\|\overline{f}\|_{L^p(B_1)}^p = \frac{\varepsilon^{2p-d}}{K^p} \int_{B_\varepsilon} |f(y)|^p \mathrm{d}y \le \varepsilon.$$

Finally, if we set $\overline{\beta}_{x_0}(x)$ to be given as

$$\overline{\beta}_{x_0}(x) := \sup_{M \in S(d)} \frac{|\overline{F}(x_0, M) - \overline{F}(x, M)|}{1 + \|M\|},$$

we get

$$\overline{\beta}_{x_0}(x) = \sup_{M \in S(d)} \frac{\left|F\left(rx_0, \frac{K}{r^2}M\right) - F\left(rx, \frac{K}{r^2}M\right)\right|}{\frac{K}{r^2} + \left\|\frac{K}{r^2}M\right\|}$$

$$\le \sup_{M \in S(d)} \frac{\left|F\left(rx_0, \frac{K}{r^2}M\right) - F\left(rx, \frac{K}{r^2}M\right)\right|}{1 + \left\|\frac{K}{r^2}M\right\|},$$

where the inequality follows from the fact that $K \ge 1$. Hence, one can always suppose u is normalized and f is arbitrarily small in $L^p(B_1)$.

Working under Assumptions 1.54 and 1.55, and the former conditions, we prove the following theorem.

Theorem 1.56 (Hölder continuity for the gradient) *Let $u \in C(\Omega)$ be a normalized L^p-viscosity solution to (1.65). Suppose Assumption 1.54 is in force for some small $0 < \theta \ll 1$, yet to be determined. In addition, suppose Assumption 1.55 holds and $F(x, 0) = 0$. Let $\alpha_0 \in (0, 1)$ be the exponent in the C^{1,α_0}-regularity for the solutions to $F(0, D^2w) = 0$. Let $\alpha \in (0, 1)$ be such that*

$$\alpha \in (0, \alpha_0) \quad and \quad \alpha \leq 1 - \frac{d}{p}.$$

Then $u \in C_{loc}^{1,\alpha}(\Omega)$. Moreover, for every $\Omega' \Subset \Omega$ there exists a constant $C > 0$ such that

$$\|u\|_{C^{1,\alpha}(\Omega')} \leq C \left(1 + \|u\|_{L^\infty(\Omega)} + \|f\|_{L^p(\Omega)}\right),$$

where $C = C(d, \lambda, \Lambda, \alpha, \mathrm{diam}(\Omega), \mathrm{dist}(\Omega', \partial\Omega))$.

As mentioned before, we notice that Ω' can be covered by a finite family of balls $B_{r/2}(x_i)$, with $i = 1, \ldots, m$, $x_i \in \Omega'$ and $r \in (0,1)$ fixed, satisfying $\overline{B_r(x_i)} \subset \Omega$. Hence, we consider (1.65) to hold in B_1 and prove an interior estimate in $B_{1/2}$.

Our next result formalizes, at the level of solutions, the former heuristic discussion on approximation methods. In truth, by supposing Assumptions 1.54 and 1.55 hold, we ensure that solutions to $F(\cdot, \cdot) = f$ and $F(0, \cdot) = 0$ can be made arbitrarily close in the L^∞-topology. We continue with an approximation lemma.

Proposition 1.57 *Let $u \in C(B_1)$ be a normalized L^p-viscosity solution to*

$$\begin{cases} F(x, D^2 u) = f & in & B_1 \\ u = g & in & \partial B_1. \end{cases} \qquad (1.67)$$

Suppose Assumptions 1.54 and 1.55 are in force, and $g \in C(\partial B_1)$. Let $\xi \in C(B_1)$ be an L^p-viscosity solution to

$$\begin{cases} F(0, D^2 \xi) = f & in & B_1 \\ \xi = g & in & \partial B_1. \end{cases} \qquad (1.68)$$

For every $\delta > 0$ there exists $\varepsilon > 0$ such that, if

$$\|\beta\|_{L^p(B_1)}, \|f\|_{L^p(B_1)} < \varepsilon,$$

then

$$\|u - \xi\|_{L^\infty(B_1)} \leq \delta.$$

Proof We argue by contradiction and suppose the statement is false. In this case, there exist sequences of functions $(u_n)_{n\in\mathbb{N}}, (\xi_n)_{n\in\mathbb{N}} \subset C(\overline{B_1})$, and $(f_n)_{n\in\mathbb{N}} \subset L^d(B_1)$, a sequence of (λ, Λ)-elliptic operators $(F_n)_{n\in\mathbb{N}}$, and a positive number $\delta_0 > 0$ such that the following hold:

(i) for every $n \in \mathbb{N}$, the functions u_n and ξ_n satisfy

$$F_n(x, D^2 u_n) = f_n \quad and \quad F_n(0, D^2 \xi_n) = 0$$

in B_1, with $u_n = \xi_n = g$ on ∂B_1;

(ii) the quantities $\beta_n(x) := \beta_n(0, x)$ and f_n are such that

$$\|\beta_n\|_{L^p(B_1)}, \|f\|_{L^p(B_1)} \leq \varepsilon_n \longrightarrow 0$$

as $n \to \infty$; but
(iii) for every $n \in \mathbb{B}$ we have

$$\|u_n - \xi_n\|_{L^\infty(B_1)} > \delta_0. \tag{1.69}$$

To produce the contradiction that completes the proof, we combine two types of results: equicontinuity for the sequences $(u_n)_{n\in\mathbb{N}}$, $(\xi_n)_{n\in\mathbb{N}}$, and $(F_n)_{n\in\mathbb{N}}$ and the stability of viscosity solutions. We proceed by splitting the proof into three steps.

Step 1. In the sequel we examine the equicontinuity of $F_n(0, \cdot)$. Because F_n is (λ, Λ)-elliptic for every $n \in \mathbb{N}$, the sequence $(F_n(0, \cdot))_{n\in\mathbb{N}}$ is uniformly bounded in the space of Lipschitz-continuous functions. As a consequence of the Arzelà–Ascoli Theorem, there exists a (λ, Λ)-elliptic operator $F_\infty(\cdot)$ such that $F_n(0, \cdot) \to F_\infty(\cdot)$ locally uniformly in $S(d)$, through a subsequence if necessary.

Notice that $\|u_n\|_{L^\infty(B_1)}$ and $\|\xi_n\|_{L^\infty(B_1)}$ are uniformly bounded. To verify the equicontinuity of $(\xi_n)_{n\in\mathbb{N}}$ we use Caffarelli and Cabré (1995, Proposition 4.14). As a result, there exists $\xi_\infty \in C(\overline{B_1})$ such that $\xi_n \to \xi_\infty$ in $C(\overline{B_1})$. It remains to check that $(u_n)_{n\in\mathbb{N}}$ is equicontinuous.

Step 2. To establish the equicontinuity of $(u_n)_{n\in\mathbb{N}}$ we resort to the existence of strong solutions in Proposition 1.16. Let $\varphi_n \in W^{2,d}_{\text{loc}}(B_1) \cap C(\overline{B_1})$ be the strong solution to

$$\mathcal{P}^-(D^2\varphi_n) = 0 \quad \text{in} \quad B_1,$$

agreeing with u_n on ∂B_1. The modulus of continuity for φ_n depends only on d, λ, Λ, $\|u_n\|_{L^\infty(B_1)}$, and the modulus of continuity of g on ∂B_1. It follows that $\overline{u_n} := u_n - \varphi_n$ is an L^p-viscosity solution to

$$F_n(x, D^2\overline{u_n}) \leq f_n$$

in B_1, vanishing at the boundary. From the estimate in Proposition 1.16 we obtain

$$\|u_n - \varphi_n\|_{L^\infty(B_1)} \leq C \|f_n\|_{L^p(B_1)} \longrightarrow 0,$$

as $n \to \infty$.

Similarly, define $\underline{u_n} := u_n - \psi_n$, where $\psi_n \in W^{2,d}_{\text{loc}}(B_1) \cap C(\overline{B_1})$ is the strong solution to

$$\mathcal{P}^+(D^2\psi_n) = 0 \quad \text{in} \quad B_1,$$

agreeing with u_n on ∂B_1. As before, $(\psi_n)_{n \in \mathbb{N}}$ is equicontinuous, with modulus of continuity depending only on universal quantities. In addition,

$$u_n - \psi_n \geq -C \|f_n\|_{L^p(B_1)}.$$

We conclude that

$$\psi_n - C \|f_n\|_{L^d(B_1)} \leq u_n \leq \varphi_n + C \|f_n\|_{L^p(B_1)},$$

with $\varphi_n = \psi_n$ on ∂B_1, for every $n \in \mathbb{N}$. This fact builds upon the inhomogeneous variant of the interior Krylov–Safonov estimate to ensure that $(u_n)_{n \in \mathbb{N}}$ is equicontinuous in $\overline{B_1}$. As a consequence, there exists $u_\infty \in C(\overline{B_1})$ such that $u_n \to u_\infty$ in $C(\overline{B_1})$, through a subsequence if necessary.

Step 3. Now we use the stability of viscosity solutions. Notice that ξ_n is a C-viscosity solution to $F_n(0, \cdot) = 0$; it follows from Caffarelli and Cabré (1995, Proposition 4.11) that $F_\infty(D^2 \xi_\infty) = 0$ in B_1, with $\xi_\infty = g$ on ∂B_1.

Now we examine u_n. For every $\varphi \in C^2(B_1)$ we notice that

$$\left| F_n(x, D^2 \varphi(x)) - f_n(x) - F_\infty(D^2 \varphi(x)) \right|$$

$$\leq \left| F_n(x, D^2 \varphi(x)) - F_n(0, D^2 \varphi(x)) \right|$$

$$+ \left| F_n(0, D^2 \varphi(x)) - F_\infty(D^2 \varphi(x)) \right| + |f_n(x)|$$

$$\leq \left| F_n(0, D^2 \varphi(x)) - F_\infty(D^2 \varphi(x)) \right| + \beta_n(x) |1 + D^2 \varphi(x| + |f_n(x)|.$$

Because the right-hand side of the former inequality converges to zero in L^p, an application of Theorem 1.17 yields $F_\infty(D^2 u_\infty) = 0$ in B_1. Because $u_\infty = \xi_\infty = g$ on ∂B_1, the uniqueness of solutions to the Dirichlet problem implies $u_\infty = \xi_\infty$. This fact contradicts (1.69) and completes the proof. ☐

Remark 1.58 (Shrinking the domain) It is usual in the literature to consider ξ as the solution to an equation holding in a slightly smaller domain – e.g., $B_{99/100} \Subset B_1$. In this case, ξ is supposed to agree with u on the boundary. The resulting proximity control in these cases also holds in the shrunk domain. Throughout the book, we also resort to this variant of approximation lemmas; however, the choice for the presentation in Proposition 1.57 intends to emphasize the fact that one can, indeed, produce an approximation result in the entire domain.

The importance of approximation lemmas in regularity theory is hardly overestimated. In fact, statements in line with Proposition 1.57 ensure that solutions to a given problem can be arbitrarily approximated by solutions to a better, more regular, equation. The core of regularity theory by approximation methods is to import this better regularity to a given problem of interest. The mechanism implementing this strategy is given by an iterative argument, which we detail next.

To prove $C^{1,\alpha}$ local estimates for the solutions to (1.65) is tantamount to prove the existence of an affine function $\ell_{x_0}(x)$ of the form

$$\ell_{x_0}(x) := a_{x_0} + \mathsf{b}_{x_0} \cdot (x_0 - x),$$

satisfying

$$\sup_{x \in B_r(x_0)} \left| u(x) - \ell_{x_0}(x) \right| \le Cr^{1+\alpha}$$

with $|\mathsf{b}_{x_0}| \le C$, uniformly in $x_0 \in B_1$, for every $0 < r \ll 1$ sufficiently small. We state this fact in the following lemma.

Lemma 1.59 (Polynomial characterization of Hölder regularity)   *Let* $u \in C(B_1)$. *For* $x_0 \in B_1$ *and* $0 < r \ll 1$, *such that* $B_r(x_0) \subset B_1$, *suppose there exists an affine function* $\ell_{x_0}(x)$ *given by*

$$\ell_{x_0}(x) := a_{x_0} + \mathsf{b}_{x_0} \cdot (x_0 - x),$$

such that

$$\sup_{x \in B_r(x_0)} \left| u(x) - \ell_{x_0}(x) \right| \le Cr^{1+\alpha}.$$

Then u *is of class* $C^{1,\alpha}$ *at* x_0. *Suppose further* $|\mathsf{b}_{x_0}| \le C$, *uniformly in* $x_0 \in B_1$, *for every* $0 < r \ll 1$. *Then* $u \in C_{loc}^{1,\alpha}(B_1)$ *and*

$$\|u\|_{C^{1,\alpha}(B_r(x_0))} \le C.$$

Proof   Start by letting $x = x_0$; it gives $u(x_0) = a_{x_0}$. Next, observe that

$$\frac{\left| u(x) - u(x_0) - \mathsf{b}_{x_0} \cdot (x - x_0) \right|}{|x - x_0|} \le C|x - x_0|^\alpha .$$

Taking the limit $x \to x_0$ in the former expression we learn that u is differentiable and $\mathsf{b}_{x_0} = Du(x_0)$. The previous inequality becomes

$$\left| u(x) - u(x_0) - Du(x_0) \cdot (x - x_0) \right| \le C|x - x_0|^{1+\alpha},$$

which implies the claim.   □

More generally, suppose $u \in C(B_1)$ and fix $x_0 \in B_1$; if there exists a polynomial $P_{x_0}(x)$ of degree k such that

$$\left| u(x) - P_{x_0}(x) \right| \le C|x - x_0|^{k+\alpha},$$

for some $\alpha \in (0,1)$, then $u \in C^{k,\alpha}$ at $x_0 \in B_1$. In the case where the partial derivatives of P_{x_0} can be uniformly bounded (with respect to x_0), then $u \in C^{k,\alpha}_{\text{loc}}(B_1)$ and its norms depend on the uniform bounds of the polynomial.

The next proposition approaches this matter through a discrete argument.

Proposition 1.60 (Oscillation control at discrete scales) *Let $u \in C(B_1)$ be a normalized L^p-viscosity solution to* (1.65). *Suppose Assumption 1.54 is in force for some small $0 < \theta \ll 1$, yet to be determined. In addition, suppose Assumption 1.55 holds and $F(x,0) = 0$. Let $\alpha_0 \in (0,1)$ be the exponent in the C^{1,α_0}-regularity for the solutions to $F(0, D^2 w) = 0$. Let $\alpha \in (0,1)$ be such that*

$$\alpha \in (0, \alpha_0) \quad and \quad \alpha \le 1 - \frac{d}{p}.$$

There exists a sequence of affine functions $(\ell_n)_{n \in \mathbb{N}}$, of the form

$$\ell_n(x) := a_n + b_n \cdot x,$$

and a number $0 < \rho \ll 1$, such that

$$\sup_{x \in B^n_\rho} |u(x) - \ell_n(x)| \le \rho^{(1+\alpha)}, \tag{1.70}$$

$$|a_n - a_{n-1}| + \rho^{n-1}|b_n - b_{n-1}| \le C\rho^{(n-1)(1+\alpha)}, \tag{1.71}$$

and

$$\left|(u - \ell_n)(\rho^n x) - (u - \ell_n)(\rho^n y)\right| \le C\rho^{n(1+\alpha)}|x - y|^\gamma, \tag{1.72}$$

for every $n \in \mathbb{N}$, every $x, y \in B_1$, and some universal constants $\gamma \in (0,1)$ and $C > 0$.

Proof We split the proof into four steps and argue by induction. The first step of the proof is the base case.

Step 1. Set $\ell_0 = \ell_{-1} \equiv 0$. Because u is a normalized solution, we have $\|u\|_{L^\infty(B_1)} \le 1$, which is (1.70) for $n = 0$. Since $\ell_0 = \ell_{-1} \equiv 0$, the condition in (1.71) is immediate. To verify (1.72), we notice that $u \in C^\gamma(B_1)$ for some $\gamma \in (0,1)$, because of Krylov–Safonov theory. It gives

$$|u(x) - u(y)| \le C|x - y|^\gamma$$

for every $x, y \in B_1$, which is precisely (1.72) for $n = 0$.

Step 2. Now we suppose the statement of the proposition has been verified for $n = k$ and examine the case $n = k + 1$. We start with the auxiliary function v_k given by

$$v_k(x) := \frac{u(\rho^k x) - \ell_n(\rho^k x)}{\rho^{k(1+\alpha)}}.$$

The induction hypothesis ensures that $\|v_k\|_{L^\infty(B_1)} \le 1$. Moreover, v_k is an L^p-viscosity solution to

$$F_k(x, D^2 v_k) = f_k \quad \text{in} \quad B_1, \tag{1.73}$$

where

$$F_k(x, M) := \rho^{k(1-\alpha)} F(\rho^k x, \rho^{k(\alpha-1)} M) \quad \text{and} \quad f_k(x) := \rho^{k(1-\alpha)} f(\rho^k x).$$

We examine $\beta_k(x)$ next. In fact,

$$\begin{aligned}
\beta_k(x) &= \sup_{M \in S(d)} \frac{|F_k(x, M) - F_k(0, M)|}{1 + \|M\|} \\
&= \sup_{M \in S(d)} \frac{\rho^{k(1-\alpha)} |F(\rho^k x, \rho^{k(\alpha-1)} M) - F(0, \rho^{k(\alpha-1)} M)|}{1 + \rho^{k(\alpha-1)} \|M\|} \\
&\le \beta(\rho^k x).
\end{aligned}$$

Hence,

$$\|\beta_k\|_{L^p(B_1)}^p \le \int_{B_1} \beta(\rho^k x)^p dx = \fint_{B_{\rho^k}} \beta(x)^p dx \le \theta^p, \tag{1.74}$$

where the last inequality follows from Assumption 1.54. When it comes to f_k, we notice that

$$\|f_k\|_{L^p(B_1)}^p = \rho^{k(1-\alpha)p} \int_{B_1} |f(\rho^k x)|^p dx = \rho^{k(1-\alpha)p} \fint_{B_{\rho^k}} |f(x)|^p dx \le C, \tag{1.75}$$

where Assumption 1.55 and the constraint on α imply the last inequality.

Step 3. Because of (1.73), (1.74), and (1.75), we learn that v_k satisfies the conditions of Proposition 1.57. Therefore, there exists $\xi \in C(B_1)$ satisfying $F_k(0, D^2 \xi) = 0$ such that

$$\|v_k - \xi\|_{L^\infty(B_1)} \le \delta.$$

In addition, because ξ is an L^p-viscosity solution to $F_k(0, D^2 \xi) = 0$, we know that $\xi \in C^{1,\alpha_0}(B_1)$ and

$$\|\xi\|_{C^{1,\alpha_0}(B_1)} \le C \|\xi\|_{L^\infty(B_1)},$$

for some universal constant $C > 0$. The regularity of ξ implies that

$$\sup_{x \in B_\rho} |v_k(x) - \xi(0) - D\xi(0) \cdot x| \leq \sup_{x \in B_\rho} (|v_k(x) - \xi(x)| + |\xi(x)$$

$$- \xi(0) - D\xi(0) \cdot x|) \qquad (1.76)$$

$$\leq \delta + C\rho^{1+\alpha_0},$$

where $C > 0$ is a universal constant. At this point, we make universal choices for $\delta > 0$ and ρ. In fact, set

$$\delta := \frac{\rho^{1+\alpha}}{2} \quad \text{and} \quad \rho := \left(\frac{1}{2C}\right)^{\frac{1}{\alpha_0 - \alpha}}. \qquad (1.77)$$

Hence (1.76) yields

$$\sup_{x \in B_\rho} \left| \frac{u(\rho^k x) - a_k - b_k \cdot (\rho^k x) - \rho^{k(1+\alpha)}\xi(0) - \rho^{k\alpha} D\xi(0) \cdot (\rho^k x)}{\rho^{k(1+\alpha)}} \right| \leq \rho^{1+\alpha}. \qquad (1.78)$$

Define ℓ_{k+1} as

$$\ell_{k+1}(x) := a_k + \rho^{k(1+\alpha)}\xi(0) + \left(b_k + \rho^{k\alpha} D\xi(0)\right) \cdot x.$$

This choice of ℓ_{k+1} builds upon (1.78) to yield

$$\sup_{x \in B_{\rho^{k+1}}} |u(x) - \ell_{k+1}(x)| \leq \rho^{(k+1)(1+\alpha)},$$

which gives (1.70) at the level $k + 1$. Moreover,

$$a_{k+1} - a_k = \rho^{k(1+\alpha)}\xi(0) \quad \text{and} \quad b_{k+1} - b_k = \rho^{k\alpha} D\xi(0),$$

which leads to (1.71), since $|\xi(0)| + |D\xi(0)| \leq C$ for some universal constant $C > 0$. To verify (1.72), we start by noticing that b_n is uniformly bounded. In fact, notice that

$$|b_1| \leq C,$$

for some universal constant $C > 0$. In addition,

$$|b_2| \leq C\rho^\alpha + C \quad \text{and} \quad |b_3| \leq C\left(1 + \rho^\alpha + \rho^{2\alpha}\right);$$

in the general case we get

$$|b_n| \leq C \sum_{m=0}^{n-1} \rho^{m\alpha},$$

where $C > 0$ is a universal constant. Now, we turn our attention to the quantity in (1.72). Because $u \in C^\gamma(B_1)$, we notice that

$$\left| (u - \ell_{k+1})(\rho^{k+1}x) - (u - \ell_{k+1})(\rho^{k+1}y) \right|$$
$$\leq C\rho^{\gamma(k+1)}|x - y|^\gamma + |b_{k+1}|\rho^{k+1}|x - y|$$
$$\leq C\rho^{(k+1)(1+\alpha)}|x - y|^\gamma,$$

and the proof is complete. $\qquad\square$

Remark 1.61 (Smallness condition on the oscillation) The (universal) choice of δ in (1.77) determines the value of $0 < \varepsilon \ll 1$ in Proposition 1.57, which is also universal. Once ε is set, the parameter $0 < \theta \ll 1$ in Assumption 1.54 is completely determined, and depends only on universal quantities.

Proof of Theorem 1.56 We split the proof into two steps.

Step 1. We start by noticing that (1.71) implies the sequences of coefficients $(a_n)_{n\in\mathbb{N}}$ and $(b_n)_{n\in\mathbb{N}}$ are Cauchy. As a result, there exist $a_\infty \in \mathbb{R}$ and $b_\infty \in \mathbb{R}^d$ such that

$$a_n \to a_\infty \quad \text{and} \quad b_n \to b_\infty,$$

as $n \to \infty$. Define $\ell_\infty(x) := a_\infty + b_\infty \cdot x$. For every $0 < r \ll 1$, we claim that

$$\sup_{x \in B_r} |u(x) - \ell_\infty(x)| \leq Cr^{1+\alpha}. \tag{1.79}$$

To verify the claim, let $n \in \mathbb{N}$ be such that $\rho^{n+1} \leq r < \rho^n$ and compute

$$\sup_{x \in B_r} |u(x) - \ell_\infty(x)| \leq \sup_{x \in B_{\rho^n}} |u(x) - \ell_\infty(x)|$$

$$\leq \sup_{x \in B_{\rho^n}} |u(x) - \ell_n(x)| + \sup_{x \in B_{\rho^n}} |\ell_n(x) - \ell_\infty(x)| \tag{1.80}$$

$$\leq \rho^{n(1+\alpha)} + \sup_{x \in B_{\rho^n}} |\ell_n(x) - \ell_\infty(x)|.$$

By observing that

$$\rho^{n(1+\alpha)} = \frac{1}{\rho}\rho^{(n+1)(1+\alpha)} < Cr^{1+\alpha}$$

and taking the limit $n \to \infty$ in (1.80), we infer the claim. Next we prove that $a_\infty = u(0)$ and $b_\infty = Du(0)$.

Step 2. Because of (1.79), we conclude that

$$|u(0) - a_\infty| \leq Cr^{1+\alpha}$$

for every $0 < r \ll 1$. Therefore it must be $a_\infty = u(0)$. To prove that $b_\infty = Du(0)$, we start by noticing that $Du(x)$ exists at $x = 0$. To that end, it suffices to verify

$$\lim_{h \to 0} \frac{|u(0 + h\theta) - u(0) - b^* \cdot (h\theta)|}{h} = 0,$$

for every $\theta \in \mathbb{S}^{d-1}$ and some $b^* \in \mathbb{R}^d$. Indeed, for $b^* = b_\infty$ it holds

$$\frac{|u(0 + h\theta) - u(0) - b^* \cdot (h\theta)|}{h} = \frac{|u(0 + h\theta) - a_\infty - b_\infty \cdot (h\theta)|}{h} \leq Ch^\alpha \to 0,$$

where the last inequality follows from (1.79). We conclude that $Du(x)$ exists at $x = 0$ and $Du(0) = b_\infty$. As a consequence, we write (1.79) as

$$\sup_{x \in B_r} |u(x) - u(0) - Du(0) \cdot x| \leq Cr^{1+\alpha},$$

which produces the result.   □

Until now, we have discussed the consequences of ellipticity and convexity on regularity theory. A remarkable feature is the effect of convexity; when combined with uniform ellipticity, it shifts the regularity regime from $C^{1,\alpha}$ to $C^{2,\alpha}$. The natural question arising in this context concerns intermediate assumptions and the respective degrees of smoothness for the solutions. In Section 1.7, we discuss a series of counterexamples clarifying this matter.

1.7 Counterexamples and Optimal Regularity

The regularity theory presented in Sections 1.2 through 1.5 can be summarized as follows. Suppose $u \in C(B_1)$ satisfies the inequality

$$\mathcal{P}^-_{\lambda,\Lambda}(D^2 u) \leq 0 \leq \mathcal{P}^+_{\lambda,\Lambda}(D^2 u) \quad \text{in} \quad \Omega$$

in the viscosity sense. Then u is locally Hölder-continuous and there exists a universal constant $C > 0$ such that

$$\|u\|_{C^\alpha(B_{1/2})} \leq C\|u\|_{L^\infty(B_1)},$$

where the Hölder exponent $\alpha \in (0, 1)$ depends only on d, λ and Λ. From the viewpoint of integral estimates, Lin's $W^{2,\varepsilon}$-regularity is also available. Suppose now that u satisfies a (λ, Λ)-elliptic equation. That is,

$$F\left(D^2u\right) = 0 \quad \text{in} \quad B_1,$$

for some (λ, Λ)-uniform elliptic operator F. In this case, $u \in C^{1,\alpha}_{\text{loc}}(B_1)$ and suitable estimates are available. If, in addition to ellipticity, we ask F to be convex on $S(d)$, then $u \in C^{2,\alpha}_{\text{loc}}(B_1)$ and estimates are available. That is, viscosity subsolutions are Hölder continuous. If we add further structure and turn subsolutions into solutions, their gradients become Hölder continuous. In the presence of even further structure (convexity), solutions are classical and their Hessians are Hölder continuous. This corpus of results suggests several fundamental questions.

The first one concerns the optimal regularity of viscosity solutions to merely uniformly elliptic equations $F = 0$. From a heuristic viewpoint, viscosity solutions to $F(D^2u) = 0$ could be of class $C^{1,1}$ or, at least, have a $C^{1, \text{Log-Lip}}$-modulus of continuity.

In other words, there is no reason a priori for the $C^{1,\alpha}$-regularity available for merely elliptic equations $F = 0$ to be optimal. This question is the underlying motivation for the following problem.

Problem 1.62 (Optimal regularity in the merely elliptic setting) Let $F : S(d) \to \mathbb{R}$ be a (λ, Λ)-elliptic operator. What is the optimal regularity for the viscosity solutions to $F\left(D^2u\right) = 0$ in B_1?

A complete answer to Problem 1.62 arises from Nadirashvili and Vlăduţ (2007, 2008, 2011); see also Nadirashvili et al. (2014). In those references, the authors argue at the intersection of nonassociative algebras and elliptic PDEs to produce several fundamental results; we highlight the following.

  (i) There exists a viscosity solution $w : B_1 \subset \mathbb{R}^{12} \to \mathbb{R}$ to a (λ, Λ)-elliptic equation $F(D^2w) = 0$ in B_1 that is not of class C^2.
 (ii) There exists a viscosity solution $v : B_1 \subset \mathbb{R}^{12} \to \mathbb{R}$ to a (λ, Λ)-elliptic equation $F(D^2v) = 0$ in the unit ball, failing to be of class $C^{1,1}$.
(iii) There exists $\alpha^* \in (0,1)$ and $u : B_1 \subset \mathbb{R}^{24} \to \mathbb{R}$ such that u solves a (λ, Λ)-elliptic equation $F(D^2v) = 0$ in B_1, $u \in C^{1,\alpha}_{\text{loc}}(B_1)$ for every $\alpha \in (0, \alpha^*)$, but $u \notin C^{1,\alpha^*}_{\text{loc}}(B_1)$.

As a conclusion, the answer to Problem 1.62 is the Hölder-continuity of the gradient. That is, in the absence of further conditions on the (λ, Λ)-elliptic operator F, solutions to $F\left(D^2u\right) = 0$ are in $C^{1,\alpha}_{\text{loc}}(B_1)$, and this level of regularity is optimal. In what follows we discuss some of the arguments in the construction of the celebrated counterexamples due to Nadirashvili and Vlăduţ (2007, 2008, 2011); our discussion is based on Nadirashvili et al. (2014).

1.7.1 A Counterexample in Dimension $d = 12$

We start by identifying an element of the quaternions with a point in the Euclidean space. In fact, for real numbers $\alpha_0, \alpha_1, \alpha_2$ and α_3, we regard

$$q_\alpha = \alpha_0 + \alpha_1 i + \alpha_2 j + \alpha_3 k \in \mathbb{H}$$

as $\alpha = (\alpha_0, \alpha_1, \alpha_2, \alpha_3) \in \mathbb{R}^4$. For $\delta \in [1, 2)$, we define $w_\delta \colon \mathbb{R}^{12} \to \mathbb{R}$ as

$$w_\delta(x) := \frac{P(x)}{\|x\|^\delta},$$

where $P \colon \mathbb{R}^4 \times \mathbb{R}^4 \times \mathbb{R}^4 \times \to \mathbb{R}$ is given by

$$\begin{aligned}
P(x, y, z) = \mathrm{Re}(q_x \cdot q_y \cdot q_z) = {} & x_0 y_0 z_0 - x_0 y_1 z_1 - x_0 y_2 z_2 - x_0 y_3 z_3 \\
& - x_1 y_0 z_1 - x_1 y_1 z_0 - x_1 y_2 z_3 + x_1 y_3 z_2 \\
& - x_2 y_0 z_2 + x_2 y_1 z_3 - x_2 y_2 z_0 - x_2 y_3 z_1 \\
& - x_3 y_0 z_3 - x_3 y_1 z_2 + x_3 y_2 z_1 - x_3 y_3 z_0.
\end{aligned}$$

A key observation concerns the role of the parameter $\delta \in [1, 2)$ in the regularity of w_δ. As δ varies in $[1, 2)$, $w_\delta \in C^{1, \alpha}$ for $\alpha \in (0, 1]$.

For $\delta = 1$ it is possible to prove that w_1 solves an equation of the form $F(D^2 w) = 0$, where F is (λ, Λ)-elliptic and smooth. It gives an example of a $C^{1,1}$ function that fails to be of class C^2. In addition, for every $\delta \in (1, 2)$ it is possible to prove that w_δ solves a *Hessian* equation driven by a (λ, Λ)-elliptic operator. We focus on the case $\delta = 1$ and discuss the general lines along which such results can be established.

To verify that w_1 solves a homogeneous (λ, Λ)-elliptic equation one uses two building blocks. First, one needs a criterion to ensure that a given function satisfies a uniformly elliptic equation. Then one must verify that w_1 satisfies such a criterion. We continue with an ellipticity criterion.

Let $w \colon \mathbb{R}^d \to \mathbb{R}$ be smooth in $\mathbb{R} \setminus \{0\}$ and suppose it is homogeneous of degree 2. As a consequence, its Hessian is homogeneous of degree 0. It is possible to associate with w a map into the space of quadratic forms on $\mathbb{R}^d$, denoted by $Q(\mathbb{R}^d)$; in fact, let $H_w \colon \mathbb{S}^{d-1} \to Q(\mathbb{R}^d)$ be given by

$$H_w(e) := D^2 w(e); \tag{1.81}$$

we notice it suffices to define it on $\mathbb{S}^{d-1}$ because $D^2 w$ is homogeneous of degree 0. It is possible to decide whether or not w is a viscosity solution to a uniformly elliptic equation based on the map H_w. We continue with a definition.

Definition 1.63 (Hessian property)   Let $w : \mathbb{R}^d \to \mathbb{R}$ be homogeneous of degree 2 on $\mathbb{R}^d$, smooth on $\mathbb{R}^d \setminus \{0\}$. Let $H_w : \mathbb{S}^{d-1} \to Q(\mathbb{R}^d)$ be defined as in (1.81). We say that w satisfies the *Hessian property* if the following conditions are met.

(i) The map $H_w : \mathbb{S}^{d-1} \to Q(\mathbb{R})$ is a smooth embedding.
(ii) Let $M, N \in H_w(\mathbb{S}^{d-1})$ and denote by $\mu_1 \geq \cdots \geq \mu_d$ the eigenvalues of $M - N$. There exists $M \geq 1$ such that

$$\frac{1}{M} < -\frac{\mu_1}{\mu_d} < M.$$

If a function w has the Hessian property, then it solves a uniformly elliptic equation. This is the content of the next proposition.

Proposition 1.64 (Ellipticity criterion)   *Suppose $w : \mathbb{R}^d \to \mathbb{R}$ satisfies the Hessian property. Then w is a viscosity solution to a uniformly elliptic equation in $\mathbb{R}^d$.*

The proof of Proposition (1.64) is detailed in Nadirashvili et al. (2014, Chapter 4.1) and we omit it here. Because of Proposition 1.64 it suffices to verify that w_1 satisfies the Hessian property to ensure that it solves a uniformly elliptic equation. This is the content of next proposition.

Proposition 1.65   *Let $w_1 : \mathbb{R}^{12} \to \mathbb{R}$ be defined as*

$$w_1(x) = \frac{P(x)}{\|x\|}.$$

Then w_1 satisfies the Hessian property.

The proof of Proposition 1.65 is the subject of Nadirashvili et al. (2014, Chapter 4.1). The interesting corollary to Proposition 1.65 is the following.

Corollary 1.66 (Nonexistence of classical solutions)   *There exists a uniformly elliptic operator $F : S(12) \to \mathbb{R}$ such that the Dirichlet problem*

$$\begin{cases} F(D^2 w) = 0 & \text{in} \quad B_1 \subset \mathbb{R}^{12} \\ w = w_1 & \text{on} \quad \mathbb{S}^{11} \end{cases}$$

has no classical solutions.

In Section 1.7.2 we revisit a question raised in Section 1.1.

1.7.2 Viscosity Inequalities and Elliptic Equations

Suppose $u \in C(B_1)$ satisfies

$$\mathcal{P}^-_{\lambda, \Lambda}(D^2 u) \leq 0 \leq \mathcal{P}^+_{\lambda, \Lambda}(D^2 u) \quad \text{in} \quad B_1. \tag{1.82}$$

An important question is whether or not there exists a (λ, Λ)-elliptic operator F such that

$$F(D^2 u) = 0 \quad \text{in} \quad B_1.$$

As mentioned before, the answer to this question is negative. It follows from Corollary 1.66. Indeed, suppose otherwise and consider w_1. Because w_1 has the Hessian property, Proposition 1.64 ensures there exists F such that $F(D^2 w_1) = 0$ in the unit ball. Hence

$$\mathcal{P}^-_{\lambda, \Lambda}\left(\frac{D^2 w_1(x + eh) - D^2 w_1(x)}{h}\right) \leq 0 \leq \mathcal{P}^+_{\lambda, \Lambda}\left(\frac{D^2 w_1(x + eh) - D^2 w_1(x)}{h}\right)$$

in B_{1-h}, for every $e \in \mathbb{S}^{11}$. Were the answer to our question affirmative, there would be a (λ, Λ)-elliptic operator G such that

$$G\left(\frac{w_1(x + eh) - w_1(x)}{h}\right) = 0,$$

in $B_{1/2}$, for every $0 < h \ll 1$. Arguing as in Section 1.4 we would conclude that $Dw_1 \in C^{1,\alpha}_{loc}(B_{1/2})$, which is impossible and leads to a contradiction.

We conclude that a function $u \in C(B_1)$ can be a viscosity solution to the inequalities in (1.82) and still there is no uniformly elliptic operator F such that $F(D^2 u) = 0$.

Bibliographical Notes

For the basics of the general theory we refer the reader to the books by Gilbarg and Trudinger (2001), Han and Lin (2011), and Caffarelli and Cabré (1995). For an account of the L^p-viscosity theory we refer the reader to the work of Caffarelli et al. (1996). See also Crandall et al. (2000) for L^p-viscosity solutions in the parabolic context.

The proof of the Krylov–Safonov theory presented here is due to Mooney (2019); we refer to that paper for a discussion on the universal dependence of Lin's exponent as well as on the modulus of continuity in the Krylov–Safonov result. A shorter proof of the Evans–Krylov result is due to Caffarelli and Silvestre (2010a); it is inspired by the authors' developments in the nonlocal setting (Caffarelli and Silvestre, 2011). For a reference on the algebraic material used in the construction of singular solutions, we refer the reader to the book by Nadirashvili et al. (2014).

For a recent account of developments pushing forward the applications of approximation methods in regularity theory, including free boundary problems, we refer to the work of Teixeira, Urbano and their collaborators; see Teixeira (2014a,b, 2016); Teixeira and Urbano (2014, 2021); Amaral and Teixeira (2015); Araújo et al. (2017, 2018, 2020); Diehl and Urbano (2020), to mention just a few.

2

Flat Solutions Are Regular

This chapter details two developments in the regularity theory for nonconvex fully nonlinear models. First, we discuss the $C^{2,\alpha}$-regularity theory for flat solutions to elliptic equations, due to Savin (2007). Second, we detail the partial regularity result due to Armstrong et al. (2012).

As discussed in Section 1.7, viscosity solutions to homogeneous uniformly elliptic equations are at most $C^{1,\alpha}$-regular, if no further conditions are imposed on the operator governing the problem. The results discussed in this chapter rely on an alternative assumption, namely, the *differentiability* of the operators.

In fact, under the assumption that F is of class C^2 in a neighborhood of $0 \in S(d)$, flat solutions to

$$F(D^2 u) = 0 \quad \text{in} \quad B_1 \tag{2.1}$$

are locally of class $C^{2,\alpha}$, where $\alpha \in (0,1)$ is universal and appropriate estimates are available. Moreover, for every $\alpha \in (0,1)$ there exists $\Sigma \subset B_1$ and $\varepsilon > 0$ such that $u \in C^{2,\alpha}(B_1 \setminus \Sigma)$, with $\dim_{\mathcal{H}}(\Sigma) < d - \varepsilon$.

In Section 2.1, we start by discussing Savin's regularity theory for flat solutions to (2.1); we refer the reader to Savin (2007).

2.1 Savin Regularity Theory in $C^{2,\alpha}$-Spaces

Let $u \in C(B_1)$ be a viscosity solution to (2.1); we say u is *flat* with respect to a constant $0 < \sigma \ll 1$ if

$$\|u\|_{L^\infty(B_1)} \le \sigma.$$

Put that way, the notion is empty. In fact, every normalized solution would be flat with respect to $\sigma := 1$. Even more delicate: given the scaling properties of (2.1), it would be possible to suppose $\|u\|_{L^\infty(B_1)} \le \sigma$ for every $0 < \sigma \ll 1$. One important aspect of this notion is the fact that σ *may depend on the operator F through structures varying with scaling arguments.* We continue with an example.

Example 2.1 (Flatness regime and scaling)   Suppose $F \in C^1(S(d))$ and denote with ω_F the modulus of continuity for DF. Suppose we are interested in a property available to the solutions of $F(D^2u) = 0$ satisfying $\|u\|_{L^\infty(B_1)} \le \sigma_F$, where $\sigma_F = \sigma_F(d, \lambda, \Lambda, \omega_F)$. It is impossible to force solutions to satisfy the σ-flatness regime through a scaling argument. Indeed, by considering

$$v(x) := \frac{\sigma u(x)}{1 + \|u\|_{L^\infty(B_1)}},$$

one notices that $\|v\|_{L^\infty(B_1)} \le \sigma$ and

$$\overline{F}(D^2v) = \frac{\sigma}{1 + \|u\|_{L^\infty(B_1)}} F\left(\frac{1 + \|u\|_{L^\infty(B_1)}}{r} D^2v\right) = 0.$$

Although the ellipticity of F is invariant under such scaling, $\omega_{\overline{F}} \neq \omega_F$ changes. In fact,

$$\left|D\overline{F}(M) - D\overline{F}(N)\right| \le \omega_F\left(\frac{1 + \|u\|_{L^\infty(B_1)}}{\sigma}|M - N|\right).$$

Because the flatness regime depends on $\omega.$, we have $\sigma_{\overline{F}} \neq \sigma_F$. Hence, by scaling the equation and fitting the initial flatness condition, we change the problem and give rise to a new flatness requirement. Since the relation between σ_F and $\sigma_{\overline{F}}$ is unknown, the strategy does not allow us to infer the desired property.

The analysis in Example 2.1 highlights the intrinsic nature of flatness. It emphasizes that this class of conditions is in fact an additional requirement on the model. Before stating the main theorem in this section, we list its main assumptions.

Assumption 2.2 (Uniform ellipticity)   The operator $F: S(d) \to \mathbb{R}$ is (λ, Λ)-uniformly elliptic; that is,

$$\lambda\|N\| \le F(M) - F(M + N) \le \Lambda\|N\|,$$

for every $M, N \in S(d)$, with $N \ge 0$. Moreover, $F(0) = 0$.

In fact, uniform ellipticity is not among the assumptions in Savin (2007). Conversely, the author imposes a (λ, Λ)-ellipticity condition in a vicinity of $0 \in S(d)$. We work under Assumption 2.2 to keep matters simpler, as our goal is to emphasize the role of differentiability and flatness in the arguments. For the sake of completeness, however, we detail an argument used in the absence of uniform ellipticity. It concerns equicontinuity for a family of functions used in the proof; see Step 2 in the proof of Theorem 2.4.

In addition to the uniform ellipticity of F, we require the operator to satisfy a differentiability condition. This is the subject of our next assumption.

Assumption 2.3 (Differentiability of F)   The operator F is of class C^2 in $S(d)$. In addition, there exists $C_F > 0$ such that $|D^2 F| \leq C_F$ in $B_{1/10} \subset S(d)$.

Under those conditions, a $C^{2,\alpha}$-regularity result is available for flat solutions to (2.1). This is the content of the next theorem.

Theorem 2.4 (Flat solutions are of class $C^{2,\alpha}$)   Let $u \in C(B_1)$ be a viscosity solution to (2.1). Suppose Assumptions 2.2 and 2.3 are in force. There exists $0 < \sigma \ll 1$ such that, if

$$\|u\|_{L^\infty(B_1)} \leq \sigma,$$

then $u \in C^{2,\alpha}_{loc}(B_1)$. Moreover, there exists a universal constant $C > 0$ satisfying

$$\|u\|_{C^{2,\alpha}(B_{1/2})} \leq C\|u\|_{L^\infty(B_1)}.$$

Finally, $\sigma = \sigma(\lambda, \Lambda, C_F)$.

The proof of Theorem 2.4 relies on two main ingredients. First we obtain a Harnack inequality and prove that the oscillation of flat solutions decays as we shrink the domain. Then, an approximation strategy completes the proof.

2.1.1 Harnack Inequality and Decay of Oscillation

We continue with a Harnack inequality for the solutions to (2.1).

Proposition 2.5 (Harnack inequality)   Let $u \in C(B_1)$ be a viscosity solution to (2.1). Suppose Assumption 2.2 holds true. There exist constants $0 < c_0 \ll 1 \ll C_0$ such that, if

$$0 \leq u(x) \leq c_0 \sigma$$

for every $x \in B_1$, then

$$u(x) \leq C_0 u(0)$$

for every $x \in B_{1/2}$.

To establish Proposition 2.5 we resort to three auxiliary lemmas, the first of which concerns a measure estimate. In fact, given a family of paraboloids of fixed opening and vertices in a closed subset of B_1, it relates the measure of the set of vertices to the measure of the contact set with respect to the solution u; compare with Lemma 1.34.

Lemma 2.6 (Measure estimate) *Let $u \in C(B_1)$ be a viscosity subsolution to (2.1). Suppose Assumption 2.2 is in force and let $a \in (0, \sigma/2]$. Fix a closed set $B \subset \overline{B}_1$; for every $y \in B$, we consider the paraboloid $P_y^a(x)$, with opening a and vertex at y, given by*

$$P_y^{-a}(x) := b_y + \frac{a}{2}|x - y|^2.$$

Slide P_y^{-a} down until it touches the graph of u from above and collect all the contact points in the set A. If $A \subset B_1 \cap \{u \leq \sigma\}$, there exists $0 < c \ll 1$ such that

$$|A| \geq c|B|.$$

Proof We split the proof into three main steps.

Step 1. Suppose at first that u is semiconvex. As a consequence, the graph of u can be touched from below by paraboloids of opening $\overline{a}$ at every point in B_1.

We claim that u is differentiable in the contact set A. It shall be clear from the fact that we can touch the graph of u from above and below with paraboloids of opening a and $\overline{a}$, respectively. At those points where Du is well-defined, we can relate the vertex of the paraboloid and the contact point by the map

$$y = x - \frac{1}{a}Du(x). \tag{2.2}$$

Our goal is to relate $|B|$ and $|A|$ through the change of variables in (2.2). To do that we examine the twice-differentiability of u.

Because u is semiconvex, the Aleksandrov Theorem ensures the existence of $Z \subset B_1$ such that u is twice-differentiable in Z and we have $|Z| = |B_1|$; see Proposition 1.25. As a result, we obtain

$$u(x) = u(z) + p(z) \cdot (x - z) + \frac{1}{2}(x - z)^T M(z)(x - z) + o\big(|x - z|^2\big), \tag{2.3}$$

for $x \in B_1$ in a vicinity of $z \in Z$. For ease of notation, we denote with $P_z(x)$ the polynomial

$$P_z(x) := u(z) + p(z) \cdot (x - z) + \frac{1}{2}(x - z)^T M(z)(x - z).$$

Step 2. We claim that for $z \in A \cap Z$ it holds that

$$-CaI \leq M(z) \leq aI, \tag{2.4}$$

where $C > 0$ is a universal constant, yet to be chosen. The second inequality in (2.4) follows from the fact that at a contact point $z \in A$, the Hessian of $u - P_y^a$ is semidefinite negative. To verify the first inequality in (2.4) we resort to a contradiction argument; i.e., for some $e \in \mathbb{S}^{d-1}$ we suppose

$$M(z) \leq -Cae \otimes e + aI.$$

First, notice that

$$P_z(x) + \frac{\varepsilon}{2}|x - z|^2 + C$$

touches the graph of u from above at some point x^* in a neighborhood of z. Because u is a subsolution to (2.1), we get

$$F(M(z) + \varepsilon I) \leq 0.$$

The uniform ellipticity of F together with the definition of $\mathcal{P}_{\lambda, \Lambda}^-$ yields

$$0 \geq F(M(z) + \varepsilon I)$$
$$\geq F((\varepsilon + a)I - Cae \otimes e) \tag{2.5}$$
$$\geq \lambda(Ca - \varepsilon - a) - \Lambda(d - 1)(\varepsilon + a);$$

by choosing $C \gg 1$ large enough, depending only on λ, Λ, and the dimension d, we obtain a contradiction.

Now we use the change of variables formula to produce

$$|B| = \int_{A \cap Z} \left| \det \left(I - \frac{1}{a}M(z) \right) \right| dx \leq C|A \cap Z|,$$

and complete the proof for semiconvex solutions.

Step 3. It remains for us to consider the general case. We consider the inf convolution u_ε and notice that the former steps yield the result for the associated contact set A_ε. Set $\varepsilon_n := 1/n$ and notice that

$$\liminf_{n \to \infty} A_{\varepsilon_n} \subset A.$$

We conclude

$$|B| \leq C \left| \liminf_{n \to \infty} A_{\varepsilon_n} \right| \leq C|A|,$$

and complete the proof. □

The next lemma provides a lower bound for the density of the contact set. Before continuing, we define the set K_a as the set of points $x \in B_1$ satisfying $u(x) \le a$ and such that there exists a paraboloid of opening a touching u from above at x.

Lemma 2.7 *Let $u \in C(B_1)$ be a viscosity subsolution to (2.1). Suppose Assumption 2.2 is in force. Let $x_0 \in B_1$ and $0 < r < 1$ be such that $\overline{B_r(x_0)} \subset B_1$ and $\overline{B_r(x_0)} \cap K_a \ne \emptyset$. There exists a universal constant $C > 0$ such that, if $a \le C^{-1}\sigma$, we have*

$$\frac{|K_{Ca} \cap B_{r/8}(x_0)|}{|B_r(x_0)|} \ge \overline{C},$$

for some $\overline{C} > 0$, universal.

Proof We start by supposing that one can find $x_1 \in B_r(x_0) \cap K_a$. In fact, if this were not the case, we could increase r slightly, by taking $r + \varepsilon$ and then letting $\varepsilon \to 0$. We consider the paraboloid $P_{y_1}(x)$ given by

$$P_{y_1}(x) := u(x_1) - \frac{a}{2}|x_1 - y_1|^2 + \frac{a}{2}|x - y_1|^2.$$

It is clear that P_{y_1} touches u from above at x_1. Also, its vertex is $y_1 \in \overline{B_1}$. Now we split the proof into two steps; first we produce an inequality relating u and P_{y_1} at a point $z \in \overline{B_{r/16}(x_0)}$. Then such inequality builds upon Lemma 2.6 to complete the proof.

Step 1. We claim there exists $z \in \overline{B_{r/16}(x_0)}$ such that

$$u(z) \ge P_{y_1}(z) - Car^2, \tag{2.6}$$

for some universal constant $C > 0$. To verify the existence of $z \in \overline{B_{r/16}(x_0)}$ such that (2.6) holds true, we introduce an auxiliary function $\varphi \colon \overline{B_1} \to \mathbb{R}$ given by

$$\varphi(x) := \begin{cases} \alpha^{-1}\big(|x|^{-\alpha} - 1\big) & \text{in } \overline{B_1} \setminus B_{1/16}, \\ \alpha^{-1}(16^\alpha - 1) & \text{in } \overline{B_{1/16}}, \end{cases}$$

where the universal constant $\alpha \gg 1$ is yet to be chosen. We add P_{y_1} to a rescaled version of $-\varphi$ to obtain $\psi \colon \overline{B_r(x_0)} \to \mathbb{R}$ given by

$$\psi(x) := P_{y_1}(x) - ar^2\varphi\left(\frac{x - x_0}{r}\right).$$

We notice that $F(D^2\psi) > 0$ in $B_r(x_0) \setminus \overline{B_{r/16}(x_0)}$. Indeed, for x in that region we get

$$
\begin{aligned}
F\big(D^2\psi(x)\big) = F\big(aI - aD^2\varphi(x)\big) \\
\geq \lambda a\big[(\alpha+1)t^{-\alpha-2} - 1\big] - \Lambda a(d-1)\big(1+t^{-\alpha-2}\big) \quad (2.7) \\
> 0,
\end{aligned}
$$

for $\alpha \gg 1$ chosen sufficiently large, where

$$
t := \frac{|x - x_0|}{r}.
$$

To make proper use of the inequality $F(D^2\psi) > 0$ in $B_r(x_0) \setminus \overline{B_{r/16}(x_0)}$, we relate the graph of ψ to the graph of u. In fact, slide down the graph of ψ until it touches the graph of u; it amounts to considering the point $z \in \overline{B_r(x_0)}$ such that

$$
(u - \psi)(z) \geq (u - \psi)(x),
$$

for every $x \in \overline{B_r(x_0)}$.

We also observe that for $x \in \partial B_r(x_0)$ the function

$$
\varphi\left(\frac{x - x_0}{r}\right) = 0;
$$

hence,

$$
u(x) \leq P_{y_1}(x) = \psi(x)
$$

on $\partial B_r(x_0)$. Moreover, since $x_1 \in B_r(x_0)$, we have

$$
u(x_1) - \psi(x_1) = P_{y_1}(x_1) - \psi(x_1) = ar^2\varphi\left(\frac{x - x_0}{r}\right) > 0;
$$

therefore, $(u - \psi)(z) > 0$. Finally notice that the maximum can not be attained at $z \in B_r(x_0) \setminus \overline{B_{r/16}(x_0)}$; were this the case, we would have $F(D^2\psi(z)) \leq 0$, which contradicts (2.7).

As a conclusion, we find that $z \in \overline{B_{r/16}(x_0)}$ and

$$
u(z) > \psi(z) \geq P_{y_1}(z) - Car^2,
$$

which is precisely (2.6).

Step 2. Now we consider a family of paraboloids of the form

$$
P_y(x) := P_{y_1}(x) + C'\frac{a}{2}|x - y|^2 + c_y,
$$

for $y \in B_{r/64}(z)$, where $C' \gg 1$ is universal, to be chosen further. The opening of P_y is $(C' + 1)a$, whereas its vertex is at

$$\frac{C'}{C'+1}y + \frac{1}{C'+1}y_1.$$

Now, observe that

$$P_{y_1}(z) + C'\frac{a}{2}|z - y|^2 + c_y \geq u(z) \geq P_{y_1}(z) - Car^2,$$

where the second inequality follows from (2.6). Hence,

$$c_y \geq -Car^2 - C'\frac{a}{2}\left(\frac{r}{64}\right)^2.$$

Finally, we conclude that the contact points of P_y and u cannot be in $B_{r/16}(z)$. In fact,

$$P_{y_1}(x) - C'\frac{a}{2}|x - y|^2 = c_y \geq P_{y_1}(x) + C'\frac{a}{2}\left(\frac{r}{32}\right)^2 - Car^2 - C'\frac{a}{2}\left(\frac{r}{64}\right)^2$$
$$> P_{y_1}(x)$$
$$\geq u(x),$$

where the strict inequality follows by choosing $C' \gg 1$ sufficiently large. We conclude the contact points must be in $B_{r/16}(z) \subset B_{r/8}(x_0)$; an application of Lemma 2.6 yields

$$\left|A_{(C'+1)a} \cap B_{r/8}(x_1)\right| \geq C|B_{r/64}(z)|,$$

which implies the result and completes the proof of the lemma.   □

We continue with the last ingredient in the proof of Proposition 2.5. It amounts to a covering lemma.

Lemma 2.8 (Covering lemma)   *Let $D_0, \dots, D_k$ be a family of closed sets satisfying*

$$D_0 \subset D_1 \subset \cdots \subset D_k \subset \cdots \subset \overline{B_{1/3}},$$

with $D_0 \neq \emptyset$. Suppose that for any $x \in B_1$ and $r \in (0,1)$ such that

$$B_r(x) \subset B_1, \quad B_{r/8}(x) \subset B_{1/3}, \quad and \quad \overline{B_r(x)} \cap D_n \neq \emptyset,$$

we have

$$\left|B_{r/8}(x) \cap D_{n+1}\right| \geq c|B_r(x)|,$$

for any $n = 0, \dots, k - 1$. Then

$$\left|B_{1/3} \setminus D_n\right| \leq (1 - \mu)^n\left|B_{1/3}\right|.$$

Proof Take D_n in the family $D_0, \ldots, D_k$ and fix $x_0 \in B_{1/3}$. Set

$$r := \text{dist}(x_0, D_n).$$

We start with the following claim: there exists $C_1 > 0$ such that

$$\left| B_{r/3}(x_0) \cap D_{n+1} \right| \geq C_1 \left| B_{1/3} \cap B_r(x_0) \right|. \tag{2.8}$$

To verify that (2.8) holds true, denote with x_1 the point

$$x_1 := x_0 - \frac{r x_0}{6|x_0|}.$$

If $y \in B_{r/6}(x_1)$, we have

$$|x_0 - y| \leq |x_0 - x_1| + |x_1 - y| \leq \frac{r}{3};$$

hence, $B_{r/6}(x_1) \subset B_{r/3}(x_0) \cap B_{1/3}$. In addition,

$$\text{dist}(x_1, D_n) \leq \text{dist}(x_1, x_0) + \text{dist}(x_0, Dn) \leq \frac{7r}{6}.$$

Together with the assumption of the lemma, the former discussion yields

$$\left| B_{r/6}(x_1) \cap D_{n+1} \right| \geq cr^d \geq c_1 \left| B_{1/3} \cap B_r(x_0) \right|$$

and establishes (2.8). Now, we consider the collection

$$\left\{ B_r(x) \mid x \in B_{1/3} \setminus D_n \text{ and } r := \text{dist}(x, D_n) \right\}$$

and take its subfamily $(B_{r_i}(x_i))_{i \in \mathbb{N}}$ such that

$$B_{1/3} \cap D_n \subset \bigcup_{i=!}^{\infty} B_{r_i}(x_i),$$

with $B_{r_i/3}(x_i) \cap B_{r_j/3}(x_j) = \emptyset$ if $i \neq j$. Hence,

$$\left| B_{1/3} \cap D_n \right| \leq \sum_{i=1}^{\infty} \left| B_{r_i}(x_i) \cap B_{1/3} \right|$$

$$\leq \frac{1}{C_1} \sum_{i=1}^{\infty} \left| B_{r_i/3}(x_i) \cap (D_{n+1} \setminus D_n) \right|$$

$$\leq \frac{1}{C_1} \left| (D_{n+1} \setminus D_n) \right|.$$

Finally, notice that

$$B_{1/3} \setminus D_{n+1} = (B_{1/3 \setminus D_{n+1}}) \setminus (D_{n+1} \setminus D_n);$$

then

$$\left|B_{1/3} \setminus D_{n+1}\right| \leq \left(1 - \frac{1}{C_1}\right)\left|B_{1/3} \setminus D_n\right|.$$

$\square$

In what follows we combine the former lemmas to detail the proof of Proposition 2.5.

Proof of Proposition 2.5   We start by assuming that $u(0) = a$, for $a < \sigma/C$; our goal is to prove that $u(x) \leq Ca$ for every $x \in B_{1/6}$, and some universal constant $C > 0$. We split the proof into three steps.

Step 1. Consider P_0^a, the polynomial of opening a and vertex at $y = 0$. Slide it down until it touches the graph of u. The choice of a implies the contact set $K_a \cap \overline{B}_{1/3} \neq \emptyset$.

For $C > 0$ as in Lemma 2.7, denote with D_k the contact set for $P_0^{aC^k}$; that is,

$$D_k := K_{aC^k} \cap \overline{B}_{1/3}.$$

It is clear that $D_0 \neq \emptyset$. In addition, if $aC^k \leq C^{-1}\sigma$, Lemma 2.7 ensures that

$$\frac{\left|D_{k+1} \cap B_{r/8}\right|}{|B_r|} \geq c.$$

Therefore, Lemma 2.8 is available for D_k and

$$\left|B_{1/3} \setminus D_k\right| \leq (1 - \mu)^k |B_{1/3}| \tag{2.9}$$

and

$$u(x) \leq aC^k \qquad \text{for} \quad x \in D_k. \tag{2.10}$$

Step 2. Next, we fix a universal constant $k_0 \in \mathbb{N}$; suppose there exists $x_k \in B_{1/6}$, for $k \geq k_0$, such that

$$u(x_k) \geq aC^{k+1}. \tag{2.11}$$

Our goal is to prove the existence of x_{k+1} such that

$$\left|x_{k+1} - x_k\right| = C_0(1 - \mu)^{k/n} =: b_k,$$

for some $C_0 >$ universal, with

$$u(x_{k+1}) \geq aC^{k+2}, \tag{2.12}$$

provided

$$4aC^{k+2}b_k^{-2} < C^{-1}\sigma. \tag{2.13}$$

To prove the existence of such x_{k+1} we argue by contradiction. Suppose that $u(x) < aC^{k+1}$ on $\partial B_{b_k}(x_k)$. We slide down the paraboloids

$$P_{k+2}(x) := 2aC^{k+2}b_k^{-2}|x - y|^2 + c_y,$$

with vertex y satisfying $|y - x_k| \leq C^{-1}b_k$, until they touch the graph of u within the cylinder

$$\{(x, z) \in \mathbb{R}^{d+1} \mid |x - x_k| \leq b_k\}.$$

We have

$$\begin{aligned} P_{k+2}(x) &\geq u(x_k) + 2aC^{k+2}b_k^{-2}\big(|x - y|^2 - |x_k - y|^2\big) \\ &\geq 2aC^k + 2aC^{k+2}b_k^{-2}|x - y|^2. \end{aligned} \tag{2.14}$$

The second inequality in (2.14) follows from the fact that $u(x_k) \geq aC^{k+1} \geq 4aC^k$ combined with

$$-2aC^{k+2}b_k^{-2}|x_k - y|^2 \leq -2aC^k.$$

Denote with K_1 the set of contact points for P_{k+1}. From (2.14) we infer two facts. First, it is clear that $u(K_1) \geq 2aC^k$. Also they belong to $B_{b_k}(x_k)$.

Under (2.13), Lemma 2.6 implies that

$$|K_1| \geq cC_0^d(1 - \mu)^k|B_1| > 2(1 - \mu)^k|B_{1/3}|.$$

By taking $k_0 \gg 1$ sufficiently large, we ensure that

$$\sum_{k=k_0}^{\infty} b_k < \frac{1}{12};$$

as a result, all the contact points are in $B_{1/12}$. From (2.9) and (2.10) we get

$$|B_{1/12}| > |B_{1/3} \setminus D_k| \leq |B_{1/3} \setminus B_{1/12}|$$

and

$$2aC^k \leq u(x) \leq aC^k,$$

if $x \in D_k$. Since these facts produce a contradiction, we conclude the existence of x_{k+1}.

Step 3. Now we suppose by contradiction that there exists $x_{k_0} \in B_{1/12}$ such that (2.11) holds true for $k = k_0$. We denote by k_* the largest value of k such that (2.13) holds. Hence,

$$4aC^{k_*+3}b_{k_*+1}^{-2} \geq C^{-1}\sigma.$$

Since

$$b_{k_*+1}^{-2} = C_0^{-2}(1-\mu)^{-2(k_*+1)/d},$$

the former inequality becomes

$$C^{-1}\sigma \le 4aC^{k_*+2}b_{k_*}^{-2}(1-\mu)^{-2/d}C_0C,$$

or, rearranging the terms,

$$\sigma C^{-3}(1-\mu)^{2/d} \le 4aC^{k_*+2}b_{k_*}^{-2}.$$

As before, we select $x_k \in B_{1/12}$, for $k = k_0, \ldots, k_*$. Now, we look for $x_{k_*+1} \in B_{1/3}$ and slide paraboloids from above in the cylinder $|x - x_{k_*+1}| \le 1/3$. On $\partial B_{1/3}$ we find

$$2aC^{k_*} + 2aC^{k_*+2}b_{k_*}^{-2}|x - y|^2 \ge 2aC^{k_*+2}b_{k_*}^{-2} \ge C^{-3}(1-\mu)^{2/d}\sigma;$$

set $c_0 := C^{-3}(1-\mu)^{2/d}$.

By assumption, $u \le c_0\sigma$; hence, the contact points must be in the interior of $B_{1/3}$. Arguing as before we reach a contradiction; it follows that $u(x) \le aC^{k_0}$ in $B_{1/6}$ and the proof is complete. □

In the next section, we prove that if u is a flat solution to $F = 0$, its L^∞-norm decays by a universal fraction when the domain changes from B_1 to $B_{1/2}$.

2.1.2 Improvement of Flatness

In what follows we suppose u is such that

$$\|u\|_{L^\infty(B_1)} \le \sigma' \le c_0\sigma \quad \text{and} \quad u(0) = 0, \tag{2.15}$$

for some $0 < \sigma' \ll 1$. We also suppose

$$-v\sigma' \le F(D^2u) \le v\sigma', \tag{2.16}$$

for some universal $0 < v \ll 1$, in the viscosity sense.

It is important to notice that Lemmas 2.6 and 2.7 remain available under (2.16), provided $a \ge v\sigma'$. In fact, it suffices to notice that (2.5) and (2.7) change accordingly when (2.16) replaces $F(D^2u) = 0$. In fact, one gets

$$\begin{aligned}
v\sigma' &\ge F(M(z) + \varepsilon I) \\
&\ge F((\varepsilon + a)I - Cae \otimes e) \\
&\ge \lambda(Ca - \varepsilon - a) - \Lambda(d-1)(\varepsilon + a) - a,
\end{aligned} \tag{2.17}$$

88 2 Flat Solutions Are Regular

instead of (2.5), and

$$F\big(D^2\psi(x)\big) = F\big(aI - aD^2\varphi(x)\big) - v\sigma'$$
$$\geq \lambda a\big[(\alpha + 1)t^{-\alpha-2} - 1\big] - \Lambda a\big(d(1 + t^{-\alpha-2})\big) - a \quad (2.18)$$
$$> v\sigma',$$

for $\alpha \gg 1$, in lieu of (2.7).

Once we have those results at our disposal, we are in a position to state a proposition on the improvement of flatness from B_1 to $B_{1/2}$.

Proposition 2.9 (Improvement of flatness) *Let $u \in C(B_1)$ be such that (2.15) and (2.16) hold true. Then*

$$\|u\|_{L^\infty(B_{1/2})} \leq (1 - v)\|u\|_{L^\infty(B_1)}.$$

Proof We aim to prove that

$$-\sigma' + v\sigma \leq u(x) \leq \sigma' - v\sigma',$$

for almost every $x \in B_{1/2}$. Suppose by contradiction that there exists $x_0 \in B_{1/2}$ such that

$$u(x_0) < -\sigma' + v\sigma'.$$

We use an argument similar to the one in the first step of the proof of Proposition 2.5 and conclude that

$$\big|\{u \geq -\sigma' + C^k v\sigma'\} \cap B_{1/2}\big| \leq (1 - \mu)^k |B_{1/2}|. \quad (2.19)$$

For $y \in B_{1/8}$, consider the paraboloids $P_y(x)$ given by

$$P - y(x) := 16\sigma'|x - y|^2 + c_y.$$

We notice that the contact points belong to

$$E := \{u \geq -\sigma'/4\} \cap B_{1/2};$$

because of Lemma 2.6 we know that

$$|E| \geq C_1,$$

for some universal $C_1 > 0$. Now we choose k such that

$$(1 - \mu)^k |B_{1/2}| \leq \frac{C_1}{2}$$

and set

$$v := \frac{1}{2C^k}.$$

The choice of ν implies that

$$E \subset \{u \geq -\sigma' + C^k \nu \sigma'\} \cap B_{1/2}.$$

As a conclusion,

$$C_1 \leq C_1/2,$$

which leads to a contradiction and completes the proof.   □

Once we have established the improvement of flatness from B_1 to $B_{1/2}$, we rescale the argument. This is the content of the next proposition.

Proposition 2.10   *Let $u \in C(B_1)$ be such that*

$$\|u\|_{L^\infty(B_1)} \leq \sigma' \leq C_0 r^2 \sigma \quad and \quad u(0) = 0, \tag{2.20}$$

with

$$-\nu r^2 \sigma' \leq F(D^2 u) \leq \nu r^2 \sigma', \tag{2.21}$$

for every $0 < r < 1$. Then

$$\|u\|_{L^\infty(B_{r/2})} \leq (1 - \nu)\sigma'.$$

Proof   The proof relies on an auxiliary function w, defined as

$$w(x) := \frac{u(rx)}{r^2}.$$

We notice that

$$\|w\|_{L^\infty(B_1)} \leq \frac{\sigma'}{r^2} \leq c_0 \sigma$$

and

$$-\nu \sigma' \leq F(D^2 w) \leq \nu \sigma'.$$

Applying Proposition 2.9 to w, with $\nu \sigma'$ replaced with $\nu r^2 \sigma'$, we get

$$\|w\|_{L^\infty(B_{1/2})} \leq (1 - \nu)\frac{\sigma'}{r^2}.$$

Rewriting the former estimate in terms of u one obtains

$$\|u\|_{L^\infty(B_{r/2})} \leq (1 - \nu)\sigma',$$

and completes the proof.   □

The next result extends the dyadic control from Proposition 2.10 to the continuous scale. We state it as a corollary.

Corollary 2.11 *Let $u \in C(B_1)$ satisfy (2.16) in the viscosity sense. If there exists $k \in \mathbb{N}$ such that*

$$\|u\|_{L^\infty(B_1)} \leq \sigma' \leq C_0 2^{-2k}\sigma \quad \text{and} \quad u(0) = 0, \tag{2.22}$$

then there exists $\rho \geq 2^{-k-1}$ and $\beta \in (0,1)$ such that

$$\|u\|_{L^\infty(B_\rho)} \leq 2\rho^\beta \sigma'.$$

Proof We split the proof into two steps; first, we prove a reduction argument.

Step 1. We claim that it suffices to prove

$$\|u\|_{L^\infty(B_{1/2^n})} \leq (1-v)^n \sigma', \tag{2.23}$$

for every $n \leq k+1$. Indeed, suppose (2.23) holds true and take $0 < s < k+1$ such that

$$\frac{1}{2^{k+1}} \leq \rho \leq \frac{1}{2^s}.$$

We have

$$\|u\|_{L^\infty(B_\rho)} \leq \|u\|_{L^\infty(B_{1/2^s})} \leq (1-v)^s \sigma' \leq \frac{2\left(2^k(1-v)^s\right)\sigma'}{2^{k+1}} \leq 2\rho^\beta \sigma',$$

where the last inequality follows from the fact that

$$2^k(1-v)^s \leq \frac{1}{\rho^{1-\beta}},$$

for some universal $\beta \in (0,1)$. Now it remains to establish (2.23).

Step 2. We prove the inequality in (2.23) by induction in n. The base case, $n = 1$, follows from Proposition 2.9. Suppose the case $n = s$ has been already verified. We prove next the case $n = s+1$. By applying Proposition 2.10 with $r = 1/2^s$ we notice that u satisfies

$$\|u\|_{L^\infty(B_{1/2^s})} \leq (1-v)^s \sigma' \leq C_0 2^{-2s}\sigma$$

and

$$-v(1-v)^s \sigma' s^{2s} \leq F(D^2 u) \leq v(1-v)^s \sigma' s^{2s}.$$

As a conclusion,

$$\|u\|_{L^\infty(B_{1/2^{s+1}})} \leq (1-v)^{s+1}\sigma',$$

which completes the proof.   □

At this point we are in a position to detail the proof of Theorem 2.4. This is the content of the next section.

2.1.3 Flat Solutions Are $C^{2,\alpha}$-Regular

In what follows we detail the proof of Theorem 2.4. We use the connection of Hölder continuity and the approximation by polynomials of certain degree; c.f. Lemma 1.59. In the following, we put forward the proof of Theorem 2.4.

Proof of Theorem 2.4   To establish the theorem we prove the existence of small constants $0 < \eta, r_0 \ll 1$ such that, if

$$\sup_{x \in B_r} \left| u(x) - a - \mathrm{b} \cdot x - x^T \frac{N}{2} x \right| \le r^{2+\alpha} \quad \text{and} \quad F(N) = 0, \qquad (2.24)$$

for $\|N\| \le \sigma/2$, one can find $a' \in \mathbb{R}$, $b' \in \mathbb{R}^d$, and $N' \in S(d)$ such that

$$\sup_{x \in B_{\eta r}} \left| u(x) - a' - b' \cdot x - x^T \frac{N'}{2} x \right| \le (\eta r)^{2+\alpha} \quad \text{and} \quad F(N') = 0.$$

$$(2.25)$$

We split the argument into four steps.

Step 1. We introduce the auxiliary function $w \in C(B_1)$, defined as

$$w(x) := \frac{u(rx) - a - \mathrm{b} \cdot (rx) - (rx)^T \frac{N}{2}(rx)}{r^{2+\alpha}};$$

our goal is to verify that

$$\sup_{x \in B_\eta} \left| w(x) - a' - b' \cdot x - x^T \frac{N'}{2} x \right| \le \eta^{2+\alpha},$$

where

$$F(N + r^\alpha N') = 0.$$

Notice that w solves

$$F\left(N + r^\alpha D^2 w\right) = 0 \quad \text{in} \quad B_1,$$

in the viscosity sense. We proceed with an equicontinuity property for w.

Step 2. We define the operator $\tilde{F}$ as

$$\tilde{F}(M) := \frac{F\left(N + r^\alpha M\right) - F(N)}{r^\alpha};$$

it is clear that $\tilde{F}$ is degenerate elliptic and uniformly elliptic in a $r^{-\alpha}$ σ-neighborhood of the origin in $S(d)$. In addition, $\tilde{F}(0) = 0$. Subtract $w(0)$ from w to produce $v := w - w(0)$ verifying

$$-C(\sigma) r^{1-\alpha} \le \tilde{F}\left(D^2 v\right) \le C(\sigma) r^{1-\alpha}.$$

Set $\sigma' := 2$ in Corollary 2.11; if

$$2 \leq C_0 2^{-2n} r^{-\alpha} \sigma \quad \text{and} \quad C(\sigma) r^{1-\alpha} \leq 2\nu,$$

we conclude that

$$\|v\|_{L^\infty(B_\rho)} \leq 4\rho^\beta,$$

for $\rho > 1/2^{k+1}$. For $x_0 \in B_{1/2}$ arbitrary, consider $v_{x_0} = w - w(x_0)$ and notice the former reasoning applies; as a consequence,

$$|w(x) - w(y)| \leq 4|x - y|^\beta,$$

for $x, y \in B_{1/2} \setminus B_{C_0 \sigma - 1/2 r^{\alpha/2}}$. Now that an equicontinuity property is available to w, we resort to a compactness argument.

Step 3. Suppose the claim associated with (2.24) and (2.25) is false. It means that regardless how small $r > 0$ is taken, there exists $\eta_0 > 0$ such that

$$\sup_{x \in B_r} \left| u(x) - a' - b' \cdot x - x^T \frac{N'}{2} x \right| \geq (\eta_0 r)^{2+\alpha} \quad \text{or} \quad F(N') \neq 0.$$

At the level of w, it means that there exist sequences $(F_n)_{n \in \mathbb{N}}$, $(N_n)_{n \in \mathbb{N}}$, $(b_n)_{n \in \mathbb{N}}$, $(a_n)_{n \in \mathbb{N}}$, $(w_n)_{n \in \mathbb{N}}$ and $(r_n)_{n \in \mathbb{N}}$ such that $r_n \to 0$,

$$F_n\left(N + r_n^\alpha D^2 w_n\right) = 0,$$
$$\|N_n\|, |b_n|, |a_n| \leq \delta/2,$$

but

$$\sup_{x \in B_{r_n}} \left| w_n(x) - a_n - b_n \cdot x - x^T \frac{N_n}{2} x \right| \geq (\eta_0 r_n)^{2+\alpha} \quad \text{or} \quad F(N') \neq 0.$$

Because F_n is uniformly elliptic, there exists F_∞ such that $F_n \to F_\infty$ and $DF_n \to DF_\infty$, locally uniformly as $n \to \infty$, through a subsequence if necessary. Because $(w_n)_{n \in \mathbb{N}}$ is equibounded and equicontinuous, there exists w_∞ such that $w_n \to w_\infty$ as $n \to \infty$, locally uniformly through a subsequence, if necessary. Finally, there exist N_∞, b_∞, and a_∞ such that

$$N_n \to N_\infty, \quad b_n \to b_\infty, \quad \text{and} \quad a_n \to a_\infty,$$

as $n \to \infty$. Denote with $DG(N)$ the matrix gathering the partial derivatives of an operator $G: S(d) \to \mathbb{R}$ with respect to the entries n_{ij} of the matrix N. We conclude that w_∞ solves

$$DF_\infty(N_\infty)D^2 w_\infty = 0 \qquad (2.26)$$

in the viscosity sense. Indeed, suppose otherwise; it is tantamount to finding $x_\infty \in B_1$ and a smooth function $\varphi(x)$ such that $\varphi(x) - \varepsilon|x - x_\infty|^2$ touches u from below at x_∞ and

$$DF_\infty(N_\infty)D^2\varphi(x_\infty) < -\varepsilon < 0.$$

The stability of minimizers implies the existence of a sequence $(x_n)_{n \in \mathbb{N}}$ such that $x_n \to x_\infty$ and φ touches w_n from below at x_k. As a conclusion,

$$0 \le \frac{F_n\big(N_n + r_n^\alpha D^2\varphi(x_n)\big) - F_n(N_n)}{r_n^\alpha} \to DF_\infty(N_\infty)D^2\varphi(x_\infty) < 0,$$

which is a contradiction, and establishes the claim.

Step 4. Because of (2.26), we infer the existence of a paraboloid

$$\overline{P}(x) := \overline{a} + \overline{b} \cdot x + \frac{x^T \overline{N} x}{2}$$

such that

$$DF_\infty(N_\infty)\overline{N} = 0,$$

and

$$\sup_{x \in B_\eta} \left| w_\infty(x) - \overline{P}(x) \right| \le \frac{\eta^{2+\alpha}}{3},$$

where $\eta = \eta(d, \lambda, \Lambda)$. Next we define the sequence $(s_n)_{n \in \mathbb{N}}$. Set

$$N_n^* := \overline{N} + s_k I$$

such that

$$F_n(N_n + r_n^\alpha N_n^*) = 0.$$

If $s_n \to 0$ as $n \to \infty$ we conclude $s_n < \eta^\alpha/3$ for $n \gg 1$ sufficiently large. As a consequence, we would have

$$\sup_{x \in B_\eta} \left| w_n(x) - \overline{a} - \overline{b} \cdot x - \frac{x^T N_n^* x}{2} \right| \le \sup_{x \in B_\eta} \left| w_n(x) - \overline{a} - \overline{b} \cdot x - \frac{x^T \overline{N} x}{2} \right| + \frac{\eta^2 s_n}{2}$$

$$\le \eta^{2+\alpha},$$

$$\qquad (2.27)$$

leading to a contradiction and completing the proof.

It remains for us to verify that $s_n \to 0$; here we use (degenerate) ellipticity. Notice that, as $n \to \infty$, we have

$$\lambda ds_n + \frac{F_n\left(N_n + r_n^\alpha \left(\overline{N} + s_k I\right)\right)}{r_n^\alpha} \leq \frac{F_n\left(N_n + r_n^\alpha \overline{N}\right)}{r_n^\alpha} \to 0.$$

It implies (2.27), leads to a contradiction and closes the argument.   □

By imposing a flatness condition on the solutions and a differentiability assumption on the operators, Theorem 2.4 ensures $C^{2,\alpha}$-regularity. The statement holds even for operator satisfying uniform ellipticity only in a vicinity of $0 \in S(d)$.

To drop the flatness condition and extend the statement for general viscosity solutions is not possible. However, it is possible to use Theorem 2.4 to examine the set where $C^{2,\alpha}$-regularity fails. This is the content of the partial regularity result discussed in Section 2.2.

2.2 The Partial Regularity Result

The regularity result due to Savin ensures local $C^{2,\alpha}$-regularity for flat solutions to $F(D^2u) = 0$, under conditions on the differentiability of F. However, when it comes to arbitrary solutions, perhaps lacking flatness, the result is no longer available.

An important development in the theory builds upon Savin's strategy and produces a *partial regularity* result for arbitrary solutions to $F(D^2u) = 0$, provided the operator is smooth enough. This development is due to Armstrong et al. (2012), and it can be stated as follows.

Theorem 2.12 (Partial regularity result)   *Let $u \in C(B_1)$ be a viscosity solution to (2.1). Suppose Assumptions 2.2 and 2.3 are in force. There exist a universal constant $\varepsilon > 0$ and a closed set $\Sigma \subset \overline{B_1}$ such that $u \in C^{2,\alpha}(B_1 \setminus \Sigma)$ for every $\alpha \in (0,1)$. Moreover,*

$$\dim_{\mathcal{H}}(\Sigma) \leq d - \varepsilon.$$

To prove Theorem 2.12 one combines three main ingredients. The first ingredient is the $W^{2,\varepsilon}$-estimates from Theorem 1.40. This estimate is applied to a linearization of $F\left(D^2u\right) = 0$; as a result we obtain a $W^{3,\varepsilon}$-estimate for the solutions. A consequence of this estimate is the existence of quadratic Taylor expansions for u, at almost every $x \in B_1$, with the third-order term uniformly controlled in $L^\varepsilon(B_1)$. In the vicinity of the points where the second-order expansion is available, Savin's theory ensures $C^{2,\alpha}$-regularity. Finally, the integrability of the third-order terms allows us to control the Hausdorff dimension of the set where $C^{2,\alpha}$-estimates fail. We continue by introducing an auxiliary function.

Let $p \in \mathbb{R}^d$ and M be a matrix. For $x \in B_1$, denote with $P^x_{p,M}(y)$ the polynomial

$$P^x_{p,M}(y) := u(x) + p \cdot (y - x) + (y - x)^T M (y - x).$$

Define $\Psi(u, B_1) \colon B_1 \to \mathbb{R}$ as

$$\Psi(u, B_1)(x) := \inf\left\{ A \geq 0 \,\middle|\, \exists\, (p, M) \text{ s.t.} \left| u(y) - P^x_{p,M}(y) \right| \leq \frac{A}{6} |x - y|^3, \forall y \in B_1 \right\}.$$

Next, we establish a result relating Ψ with the function Θ introduced in Section 1.3.

Proposition 2.13 *Let $u \in C^1(B_1)$. Then*

$$\Psi(u, B_1)(x) \leq \left[\sum_{i=1}^{d} \left(\Theta(u_{x_i}, B_1)(x) \right)^2 \right]^{\frac{1}{2}},$$

for every $x \in B_1$.

Proof Let $x \in B_1$ and denote by $A_i \geq 0$ constants satisfying

$$A_i := \Theta(u_{x_i}, B_1)(x).$$

From the definition of Θ we infer the existence of $p^i \in \mathbb{R}^d$ such that

$$\left| u_{x_i}(y) - u_{x_i}(x) + p^i(x - y) \right| \leq \frac{A_i}{2} |x - y|^2, \qquad (2.28)$$

for every $y \in B_1$. For any matrix M, we have

$$u(y) - u(x) + Du(x) \cdot (x - y) + (x - y)^T M (x - y)$$

$$= (y - x) \cdot \int_0^1 Du(x + \tau(y - x)) - Du(x) + 2\tau M(y - x) d\tau. \qquad (2.29)$$

Define $M = (m_{i,j})^d_{i,j=1}$ as

$$m_{i,j} := \frac{1}{2} p^i_j,$$

and use (2.28) with (2.29) we obtain

$$\int_0^1 \left| u_{x_i}(x + \tau(y - x)) - u_{x_i}(x) + \tau p^i_j \cdot (y - x) \right| d\tau \leq \frac{A_i}{2} \tau^2 |x - y|^2.$$

Let $A := (A_1, \ldots, A_d)$; the former inequality then builds upon (2.29) to produce

$$
\begin{aligned}
&\left| u(y) - u(x) + Du(x) \cdot (x - y) + (x - y)^T M(x - y) \right| \\
&\leq (y - x) \cdot \int_0^1 \frac{A\tau^2}{2} |x - y|^2 d\tau \\
&\leq \frac{1}{6} |A| |x - y|^3.
\end{aligned}
$$

It follows that

$$
\Psi(u, B_1)(x) \leq |A| = \left[\sum_{i=1}^d \left(\Theta(u_{x_i}, B_1)(x) \right)^2 \right]^{\frac{1}{2}},
$$

which completes the proof.   □

Next we study a measure decay rate for the super level sets of Ψ.

Corollary 2.14 (Measure decay for Ψ) *Let $u \in C(B_1)$ be a normalized viscosity solution to (2.1). There exist universal constants $C > 0$ and $0 < \varepsilon \ll 1$ such that*

$$
\left| \{ x \in B_{1/2} \mid \Psi(u, B_1)(x) > \tau \} \right| \leq C\tau^{-\varepsilon},
$$

for every $\tau > 1$.

Proof Because u solves (2.1), Theorem 1.46 ensures $u \in C^{1,\alpha}_{\mathrm{loc}}(B_1)$. In particular, it is of class C^1. Moreover, for every $e \in \mathbb{S}^{d-1}$, the directional derivative $u_e := e \cdot Du$ satisfies

$$
\mathcal{P}^+ \left(D^2 u_e \right) \geq 0 \quad \text{in} \quad B_1.
$$

Hence, Lin's integral estimates, as stated in Theorem 1.40, are available for u_e. As a consequence, there are universal constants $C > 0$ and $\varepsilon > 0$ such that

$$
\left| \{ x \in B_{1/2} \mid \Theta(u_e, B_1)(x) > \tau \} \right| \leq C\tau^{-\varepsilon}.
$$

Because of Proposition 2.13, we know that $\Psi(u, B_1)(x) \leq \Theta(u_{x_i}, B_1)(x)$, for every $x \in B_1$. Therefore,

$$
\{ x \in B_{1/2} \mid \Psi(u, B_1)(x) > \tau \} \subset \{ x \in B_{1/2} \mid \Theta(u_e, B_1)(x) > \tau \},
$$

which completes the proof.   □

Proposition 2.15 *Let $u \in C(B_1)$ be a viscosity solution to (2.1). Suppose Assumptions 2.2 and 2.3 are in force. Fix $\alpha \in (0, 1)$; it is arbitrary. There exists a constant $\delta_\alpha > 0$ such that, if for every $y \in B_{1/2}$ and $0 < r \ll 1$ we have*

$$\left\{\Psi(u, B_1) \le r^{-1}\delta_\alpha\right\} \cap B_r(y) \ne \emptyset,$$

then $u \in C^{2,\alpha}(B_r(y))$.

Proof Take $y \in B_{1/2}$ and fix $0 < r \ll 1$. Suppose

$$\Psi(u, B_1)(z) \le r^{-1}\delta,$$

for every $z \in B_r(y)$, where $\delta > 0$ is yet to be determined. The definition of Ψ ensures the existence of $(p, M) \in \mathbb{R}^d \times S(d)$ such that

$$\left|u(x) - u(z) + p \cdot (z - x) + (z - x)^T M(z - x)\right| \le \frac{r^{-1}\delta}{6}|z - x|^3. \quad (2.30)$$

We proceed by noticing that

$$\varphi(y) := u(z) - p \cdot (z - x) - (z - x)^T M(z - x)$$

touches the graph of u at $y = x$. In addition,

$$\phi(y) := u(x) + p(x - y) + (x - y)^T M(x - y)$$

touches the graph of u at $y = x$. As a consequence, $F(-M) = 0$.
  Next define $v \colon B_1 \to \mathbb{R}$ as

$$v(x) := \frac{u(z + 4rx) - u(z) - 4rp \cdot x + 16r^2 x^T M x}{16r^2}.$$

Our goal is to apply Theorem 2.4 to v. Because of (2.30) we have

$$\sup_{x \in B_1} |v(x)| \le \frac{2\delta}{3}.$$

At this point we define $\overline{F}(N) := F(N - M)$. It is immediate that $\overline{F}$ inherits the uniform ellipticity of F. Also, $\overline{F}$ satisfies Assumption 2.3 with the same modulus of continuity and bounds. Finally, $\overline{F}(0) = F(-M) = 0$. We proceed by observing that $\overline{F}(D^2 v) = 0$ in B_1.
  Denote with $\sigma > 0$ the flatness regime in Theorem 2.4. By choosing

$$\delta := 3e^{-1}\sigma,$$

we conclude $v \in C^{2,\alpha}(B_{1/2})$. Since

$$D^2 v(x) = D^2 u(z + 4rx),$$

we conclude $u \in C^{2,\alpha}(B_{2r}(z))$; because $B_r(y) \subset B_{2r}(z)$, the proof is complete. □

  Next we detail the proof of Theorem 2.12.

Proof of Theorem 2.12   We split the proof into three steps. The first one accounts for simplifications we use in what follows.

Step 1. We start by fixing $\alpha \in (0,1)$. Moreover, a covering argument allows us to consider $\Omega = B_1$ and to prove that $u \in C^{2,\alpha}_{\text{loc}}(B_{1/2} \backslash \Sigma)$, where $\dim_{\mathcal{H}}(\Sigma) \leq d - \varepsilon$. We also suppose $\|u\|_{L^\infty(B_1)} \leq 1$. In fact, consider $\bar{u}(x) := (\|u\|_{L^\infty(B_1)})^{-1} u(x)$. Then

$$\frac{1}{\|u\|_{L^\infty(B_1)}} F\left(\|u\|_{L^\infty(B_1)} D^2 \bar{u}\right) = \frac{1}{\|u\|_{L^\infty(B_1)}} F\left(D^2 u\right) = 0.$$

To complete the argument we remark that the operator

$$\frac{1}{\|u\|_{L^\infty(B_1)}} F\left(\|u\|_{L^\infty(B_1)} M\right)$$

has the same ellipticity constants as F, although its derivatives have distinct moduli of continuity. Because the exponent $\varepsilon > 0$ does not depend on the moduli of continuity of $D^2 F$, the argument is in force.

Step 2. Define the set Σ as

$$\Sigma := \left\{ x \in B_{1/2} \mid u \notin C^{2,\alpha}(B_r(x)) \; \forall \, r > 0 \right\}.$$

We claim that Σ is closed. To verify this claim, consider $y \notin \Sigma$. There exists $r > 0$ such that $u \in C^{2,\alpha}(B_r(y))$. It is clear that $B_{r/100}(y) \not\subset \Sigma$ and the claim follows. Since Σ is also bounded, it is compact.

As a consequence, for $0 < r \ll 1$ small enough, there exists a finite collection of balls $B_r(x_1), \ldots, B_r(x_m)$, such that

$$\Sigma \subset \bigcup_{i=1}^{m} B_r(x_i).$$

We apply the contrapositive of Proposition 2.15 to each $x_i \in \Sigma$ to conclude that, for every $y \in B_r(x_i)$, we get

$$\Psi(u, B_1)(y) > r^{-1}\delta.$$

Step 3. Corollary 2.14 implies

$$Cr^d \leq C|B_r(x_i)| \leq C\left|\left\{\Psi(u, B_1) > r^{-1}\delta\right\}\right| \leq Cr^\varepsilon.$$

Hence,

$$\sum_{i=1}^{m} \left|B_r(X_i)\right|^{d-\varepsilon} \leq C,$$

for some universal constant $C > 0$. We conclude $\mathcal{H}^{d-\varepsilon}(\Sigma) \leq C$ and complete the proof.   □

Bibliographical Notes

We have not emphasized the lack of uniformly ellipticity in Savin's argument. The reader is referred to his original paper (Savin, 2007) where the precise assumptions are stated. We also suggest Caffarelli's work on the Monge–Ampère equations for an account on important ideas in the degenerate context; see, for instance, Caffarelli (1990, 1991). Armstrong and Tran (2015) and Capuzzo Dolcetta et al. (2010) have also examined degenerate elliptic Hamilton–Jacobi equations, with important result on the regularity of the solutions. See also the work of Oberman and Silvestre (2011) and Birindelli et al. (2018). As the partial regularity result suggests, some background in geometric measure theory is useful when approaching this literature. In that direction we suggest the book by Evans and Gariepy (2015) and the book by Maggi (2012).

3

The Recession Strategy

In this chapter we develop the notion of a *recession operator* in connection with regularity theory. It amounts to an asymptotic method seeking to import regularity from the ends of the space of symmetric matrices. In particular, it allows us to displace our assumptions from the original problem to an auxiliary one operating at the infinity of $S(d)$. A by-product of this technology is in the realm of *weak regularity theory*, which we describe further.

3.1 The Recession Function

Given a fully nonlinear operator $F: S(d) \to \mathbb{R}$, we can define the recession function associated with F. This object, denoted by F^*, is also a (λ, Λ)-elliptic operator, and accounts for the properties of F at the ends of $S(d)$. The recession operator is useful in the context of approximation methods and regularity theory.

One aims to transmit information from the equation governed by F^* to the original one. In brief, one supposes F^* satisfies a set of assumptions and examines its consequences on the solutions to $F = 0$. In what follows we detail the basics associated with this object. We begin with the formal definition of a *recession operator*.

Definition 3.1 (Recession operator)   Let $F: S(d) \to \mathbb{R}$ be a (λ, Λ)-elliptic operator. For $\mu > 0$, set

$$F_\mu(M) := \mu F\left(\mu^{-1} M\right).$$

The recession function associated with F is the operator $F^*: S(d) \to \mathbb{R}$ defined by

$$F^*(M) := \lim_{\mu \to 0} F_\mu(M).$$

Remark 3.2 Throughout the text, we also refer to F^* as the *recession function* or *recession profile*.

Remark 3.3 (Sending $\mu \to 0$) Because F_μ is (λ, Λ)-elliptic for every $\mu > 0$, the family $(F_\mu)_{\mu>0}$ is uniformly Lipschitz continuous. As a consequence, F_μ converges locally uniformly as $\mu \to 0$, through a subsequence if necessary. Of course, these subsequential limits might not coincide. At this point we have an alternative. First, we can suppose these subsequential limits to be the same; it amounts to requiring F^* to be unique. Alternatively, we can suppose that every subsequential limit meets the assumptions involved in each particular context; in this case, *every recession function* associated with F would satisfy the conditions required in the analysis.

An interesting observation concerns the ellipticity of F^*; it is the same as the uniform ellipticity of F. If the latter is a (λ, Λ)-elliptic operator, so is F^*. Indeed, for given $M, N \in S(d)$, with $N \geq 0$, we have

$$F^*(M + N) - F^*(M) = \mu F\left(\mu^{-1}(M + N)\right) - \mu F\left(\mu^{-1}M\right)$$
$$\leq \mu\left(\Lambda\|\mu^{-1}N\|\right)$$
$$= \Lambda \|N\|.$$

The remaining inequality follows similarly. Therefore, the recession function associated with a particular fully nonlinear elliptic operator yields an operator in the same (λ, Λ)-ellipticity class.

The recession operator plays the role of the fixed-coefficients counterpart of $F(x, M)$ in Section 1.6. To properly connect F and F^* through an approximation strategy, we start by establishing a few properties of the recession profile. We start with the homogeneity of this operator.

Lemma 3.4 (Positive homogeneity of degree 1) *Let $F: S(d) \to \mathbb{R}$ be a (λ, Λ)-elliptic operator and F^* be its recession profile. Suppose F^* is unique. Then it is positively homogeneous of degree 1.*

Proof Let $\rho > 0$ be fixed; we compare $F^*(\rho M)$ and $\rho F^*(M)$. The definition of recession function yields

$$\left|F^*(\rho M) - \rho F^*(M)\right| \leq \left|F^*(\rho M) - F_\mu(\rho M)\right| + \left|F_\mu(\rho M) - \rho F^*(M)\right|. \tag{3.1}$$

For every $\delta > 0$, there exists $\varepsilon > 0$ such that $\mu \in (0, \varepsilon)$ implies

$$\left|F^*(\rho M) - F_\mu(\rho M)\right| \leq \delta.$$

In addition, notice that

$$\left|F_\mu(\rho M) - \rho F^*(M)\right| = \rho \left|F_{\mu\rho^{-1}}(M) - F^*(M)\right|;$$

the uniqueness of the recession function yields

$$\rho \left|F_{\mu\rho^{-1}}(M) - F^*(M)\right| \to 0$$

as $\mu \to 0$. Combining the former computations we produce

$$\left|F^*(\rho M) - \rho F^*(M)\right| \le \varepsilon^*,$$

for arbitrarily small ε^* and finish the proof.   □

Next, we use Lemma 3.4 to study the uniform convergence of F_μ to F^*, as $\mu \to 0$. Such a result is instrumental when studying regularity theory by approximation methods.

Proposition 3.5 (Uniform convergence) *Let $F: S(d) \to \mathbb{R}$ be a (λ, Λ)-elliptic operator and let F^* be its recession profile. Suppose F^* is homogeneous of degree 1. Then F_μ converges locally uniformly to F^*. Moreover, for every $\delta > 0$ there exists $\varepsilon > 0$ so that*

$$\left|F_\mu(M) - F^*(M)\right| \le \varepsilon(1 + \|M\|), \tag{3.2}$$

provided $\mu \le \delta$.

Proof First we establish the uniform convergence; then we prove the estimate in (3.2). We split the argument into two steps.

Step 1. Because F_μ is (λ, Λ)-elliptic, the family $(F_\mu)_{\mu>0}$ is uniformly Lipschitz continuous in $S(d)$. An application of the Arzelà–Ascoli Theorem, implies that F_μ converges locally uniformly, through a subsequence if necessary. The definition of F^* implies that $F_\mu(M)$ converges pointwise to $F^*(M)$, for every $M \in S(d)$. Therefore, every subsequential limit F_{μ_j} must coincide, as $j \to \infty$. We conclude that F_μ converges uniformly locally to F^*.

Step 2. As for the estimate in (3.2), we consider two cases. Suppose first that $\|M\| \le 1$. Here, (3.2) follows from the local uniform convergence of F_μ. If, otherwise, $\|M\| > 1$, consider

$$\mu_M := \frac{\mu}{\|M\|}.$$

By assumption, F^* is positively homogeneous of degree 1. Then

$$\frac{1}{\|M\|}\left|F_\mu(M) - F^*(M)\right| = \left|F_{\mu_M}\left(\frac{M}{\|M\|}\right) - F^*\left(\frac{M}{\|M\|}\right)\right| \to 0, \tag{3.3}$$

as $\mu_M \to 0$, where we have used the first case. From (3.3), we get

$$\left| F_\mu(M) - F^*(M) \right| \leq \varepsilon \|M\|,$$

for $\mu \ll 1$ and complete the proof.   □

The uniform convergence ensured by Proposition 3.5 implies that F_μ and F^* can be taken arbitrarily close in the uniform topology, provided $\mu \ll 1$ satisfies a required smallness regime.

Before we continue, we present a few examples and compute the recession operator explicitly.

Example 3.6 Fix $q \in 2\mathbb{N} + 1$ and consider the operator

$$F_q(M) := \sum_{i=1}^{d} \left(1 + e_i^q \right)^{\frac{1}{q}},$$

where $(e_i)_{i=1}^{d}$ are the eigenvalues of the matrix M. We emphasize that F_q is neither convex nor concave. We compute

$$\mu F_q\left(\mu^{-1} M \right) = \mu^{q/q} \sum_{i=1}^{d} \left(1 + \mu^{-q} e_i^q \right)^{\frac{1}{q}} = \sum_{i=1}^{d} \left(\mu^q + e_i^q \right)^{\frac{1}{q}}.$$

Therefore,

$$F_q^*(M) = \lim_{\mu \downarrow 0} \sum_{i=1}^{d} \left(\mu^q + e_i^q \right)^{\frac{1}{q}} = \mathrm{Tr}(M).$$

This example shows that the recession operator relates F_q to the Laplacian. Moreover, if we are interested in ellipticity-invariant (or universal) properties of F_q, it suffices to examine the case of the Laplacian operator.

Our next example appears in differential geometry. It is a perturbation of the special Lagrangian equation.

Example 3.7 Consider the special Lagrangian operator

$$F(M) := \sum_{i=1}^{d} \arctan\left(1 + e_i \right) + \alpha_i e_i,$$

where $(\alpha_i)_{i=1}^{d}$ are real numbers. A straightforward computation yields

$$F^*(M) = \sum_{i=1}^{d} \alpha_i e_i;$$

i.e., the operator under analysis relates to a perturbation of the Laplacian.

Example 3.8 (The log-Monge–Ampère operator)   The log-Monge–Ampère operator is given by

$$F(M) := \ln[\det(M)].$$

Consider uniformly convex solutions and suppose the eigenvalues of M are strictly above 1. Fix $\alpha_i \in \mathbb{R}$, with $|\alpha_i| \ll 1$ sufficiently small; then study the perturbed problem

$$F_\alpha(M) := \ln[\det(M)] + \sum_{i=1}^{d} \alpha_i \lambda_i.$$

Because $e_i > 1$, the sublinearity of the logarithm yields

$$\mu \left[\ln\left[\det\left(\mu^{-1}M\right)\right] + \sum_{i=1}^{d} \alpha_i \mu^{-1}\lambda_i \right] \leq C(d)\sqrt{\mu} + \sum_{i=1}^{d} \alpha_i \lambda_i;$$

hence,

$$F_\alpha^*(M) = \sum_{i=1}^{d} \alpha_i \lambda_i.$$

Section 3.2 resorts to the recession operator to produce regularity results in Sobolev spaces.

3.2 Applications to Regularity Theory in Sobolev Spaces

In this section we consider the equation

$$F(D^2u) = f \quad \text{in} \quad B_1, \tag{3.4}$$

where F is a (λ, Λ)-elliptic operator with $F(0) = 0$, and $f \in L^p(B_1)$, with $p > d$. By imposing a condition on the regularity of the solutions to $F^*(D^2v) = 0$, we prove that solutions to (3.4) are in $W^{2,p}_{loc}(B_1)$ with the appropriate estimates. This is the content of the next theorem.

Theorem 3.9 ($W^{2,p}$-regularity)   *Let $u \in C(B_1)$ be a normalized viscosity solution to (3.4). Suppose F is (λ, Λ)-elliptic $F(0) = 0$, and F^* is convex. Suppose further $f \in L^p(B_1)$, for some $p > d$. Then $u \in W^{2,p}_{loc}(B_1)$ and*

$$\|u\|_{W^{2,p}(B_{1/2})} \leq C \left(\|u\|_{L^\infty(B_1)} + \|f\|_{L^p(B_1)} \right),$$

where $C > 0$ is a universal constant.

Remark 3.10 In Theorem 3.9, we require F^* to be a convex operator. This condition is imposed on the recession operator merely to take advantage of the Evans–Krylov theory, available for the solutions to

$$F^*(D^2 v) = 0 \quad \text{in} \quad B_1; \tag{3.5}$$

see Section 1.5. However, the proof of Theorem 3.9 depends on a much less strict assumption. In fact, it suffices to have $W^{2,q}$-estimates available for $F = 0$, to ensure regularity in $W^{2,p}$-spaces, for $d < p < q$; see Li and Zhang (2015). Caffarelli and Cabré (1995) require $C^{1,1}$-estimates for the solutions to (3.5) instead of convexity of the operator. That is, $v \in C^2(B_1) \cap C(\overline{B}_1)$ and there exists $C > 0$, universal, so that

$$\|v\|_{C^{1,1}(B_{9/10})} \le C.$$

Theorem 3.9 appeared in Pimentel and Teixeira (2016), as a recession-based variant of the result in Caffarelli (1989); see also Caffarelli and Cabré (1995, Chapter 7). It can be framed as a Calderón–Zygmund estimate. The analysis involves second-order polynomials touching the graph of u from above and below. We introduce notation and definitions in the sequel.

A convex quadratic polynomial of opening $M > 0$ is a function $P_M^+ : B_1 \to \mathbb{R}$ of the form

$$P_M^+(x) := \ell(x) + M\frac{|x|^2}{2},$$

where $\ell : B_1 \to \mathbb{R}$ is an affine function. Similarly, a concave quadratic polynomial $P_M^- : B_1 \to \mathbb{R}$ of opening $M > 0$ has the form

$$P_M^-(x) := \ell(x) - M\frac{|x|^2}{2};$$

we switch from convex to concave polynomials depending on the nature of the analysis. That is, when touching the graph of solutions from above, we resort to convex paraboloids, whereas touching from below requires concave ones. When there is no room for confusion, we drop the superscript and write only P_M.

Definition 3.11 Let $u \in C(B_1)$. For $M > 0$ and $H \subset B_1$, we define

$$\underline{G}_M(u, H) := \left\{ x \in H \mid \exists P_M^- \text{ touching } u \text{ from below at } x \right\}$$

and

$$\overline{G}_M(u, H) := \left\{ x \in H \mid \exists P_M^+ \text{ touching } u \text{ from above at } x \right\}.$$

We set

$$\underline{A}_M(u, H) := H \setminus \underline{G}_M(u, H) \qquad \text{and} \qquad \overline{A}_M(u, H) := H \setminus \overline{G}_M(u, H).$$

Also,

$$G_M(u, H) := \underline{G}_M(u, H) \cap \overline{G}_M(u, H)$$

and

$$A_M(u, H) := \underline{A}_M(u, H) \cup \overline{A}_M(u, H).$$

Finally, we define the $\Theta \colon H \to \mathbb{R}$ as

$$\Theta(x) := \inf \{ M > 0 \mid x \in G_M(u, H) \}.$$

Compare Θ with the function $\underline{\Theta}$ in Section 2.2. We continue with a variant of Theorem 1.40 for the inhomogeneous case.

Proposition 3.12 (Inhomogeneous $W^{2,\delta}$-estimates) *Let $u \in C(B_1)$ be a normalized viscosity solution to (3.4). Suppose F is (λ, Λ)-elliptic $F(0) = 0$, and F^* is convex. Suppose further $f \in L^p(B_1)$, for some $p > d$. There exist $\delta > 0$ and $C > 0$, universal constants, such that*

$$\int_{B_1} |\Theta(x)|^\delta \, dx \leq C \left(\|u\|_{L^\infty(B_1)} + \|f\|_{L^d(B_1)} \right).$$

To adapt the proof of Theorem 1.40 to the inhomogeneous case is a neat exercise; we refer the interested reader to the routine in Caffarelli and Cabré (1995, Lemmas 7.7 and 7.8) for the required modifications. The recession strategy builds upon Proposition 3.12; in fact, the regularity available for $F^* = 0$ accelerates the decay rate for the measure of the sets A_M. We recall a fundamental result.

Proposition 3.13 *Let $g \geq 0$ be a measurable function on B_1 and denote by μ_g its distribution function*

$$\mu_g(t) = |\{ x \in B_1 \mid g(x) > t \}|, \qquad t > 0.$$

Fix $\zeta > 0$ and $M > 1$. For $p > 0$, we have

$$g \in L^p(B_1) \qquad \Longleftrightarrow \qquad \sum_{k=1}^{\infty} M^{pk} \mu_g(\zeta M^k) =: S < \infty.$$

Moreover, for some $C = C(\zeta, M, p)$, we have

$$C^{-1} S \leq \|g\|_{L^p(B_1)}^p \leq C(1 + S).$$

For more on Proposition 3.13, we refer the reader to Caffarelli and Cabré (1995, Lemma 7.3). Proposition 3.13 implies that $D^2 u \in L^p(B_{1/2})$ is equivalent to the summability of

$$\sum_{k=1}^{\infty} M^{pk} \left| A_{M^k}(u, B_{1/2}) \right|,$$

for some fixes M.

Here we use the recession strategy. The estimates available for $F^* = 0$ set a competing inequality: When the Hessian of the solutions to (3.4) grows, the recession starts to govern the problem. Because it is convex, the norm of the Hessian decreases and the original operator resumes driving the equation. This process prevents the Hessian from blowing up in an L^p-sense. We connect $F = f$ and $F^* = 0$ through an approximation lemma.

Proposition 3.14 *Let $u \in C(B_1)$ be a viscosity solution to*

$$F_\mu\left(D^2 u\right) = f \quad in \quad B_1,$$

where F is (λ, Λ)-elliptic $F(0) = 0$. Suppose $f \in L^p(B_1)$, for some $p > d$. Suppose further that F^* is convex. For every $\delta > 0$, there exists $\varepsilon > 0$ such that if

$$\mu + \|f\|_{L^p(B_1)} \leq \varepsilon,$$

one can find $h \in C_{loc}^{2,\alpha}(B_1)$, with

$$\|h\|_{C^{2,\alpha}(B_{9/10})} \leq C \|h\|_{L^\infty(B_1)},$$

satisfying

$$\|u - h\|_{L^\infty(B_{9/10})} \leq \delta.$$

Here $C > 0$ and $\alpha \in (0, 1)$ are universal constants.

Proof As usual, we argue by way of contradiction and use a compactness argument; we split the proof into two steps.

Step 1. Suppose the statement of the proposition is false. In this case, there exists a sequence of real numbers $(\mu_n)_{n \in \mathbb{N}}$ and sequences of functions $(u_n)_{n \in \mathbb{N}}$ and $(f_n)_{n \in \mathbb{N}}$ such that

$$\mu_n \to 0, \quad \|f_n\|_{L^p(B_1)} \to 0,$$

and

$$F_{\mu_n}\left(D^2 u_n\right) = f_n \quad in \quad B_1,$$

but, for every function $h \in C_{\text{loc}}^{2,\alpha}(B_{9/10})$, we have

$$\|u_n - h\|_{L^\infty(B_{9/10})} \geq \delta_0,$$

for some $\delta_0 > 0$.

Step 2. Because of the regularity theory available for $F_{\mu_n} = f_n$, we have that $(u_n)_{n \in \mathbb{N}}$ is equibounded in $C_{\text{loc}}^{1,\alpha}(B_1)$, for some $\alpha \in (0,1)$. Therefore, there exists $u_\infty \in C_{\text{loc}}^{1,\frac{\alpha}{2}}(B_{9/10})$ such that $u_n \to u_\infty$, through a subsequence if necessary, in the $C^{1,\frac{\alpha}{2}}$-topology. The stability of viscosity solutions, Theorem 1.17, implies that

$$F^*(D^2 u_\infty) = 0 \quad \text{in} \quad B_{9/10}.$$

Moreover, the Evans–Krylov theory yields $u_\infty \in C_{\text{loc}}^{2,\alpha}(B_{9/10})$ and

$$\|u_n - u_\infty\|_{L^\infty(B_{9/10})} \to 0,$$

as $n \to \infty$. By taking $h \equiv u_\infty$, we get a contradiction and complete the proof.   □

Next, we combine Proposition 3.14 with Proposition 3.12 to control the measure of the sets

$$G_M(u, B_1) \cap Q_1.$$

Lemma 3.15 *Let $u \in C(B_1)$ be a viscosity solution to (3.4) and suppose*

$$-|x|^2 \leq u(x) \leq |x|^2 \quad \text{in} \quad B_1 \setminus B_{3/4}.$$

Under the assumptions of Proposition 3.14, there exists $M > 0$, depending only on the dimension, and $\rho \in (0,1)$ such that

$$|G_M(u, B_1) \cap Q_1| \geq 1 - \rho.$$

Proof Let h be the approximating function whose existence is ensured by Proposition 3.14 and consider its restriction to $B_{1/2}$. Extend h outside $\overline{B}_{1/2}$ continuously in such a way that

$$h \equiv u \quad \text{in} \quad B_1 \setminus B_{3/4}$$

and

$$\|u - h\|_{L^\infty(B_1)} = \|u - h\|_{L^\infty(B_{3/4})}.$$

We conclude

$$-2 - |x|^2 \leq h(x) \leq 2 + |x|^2 \quad \text{in} \quad B_1 \setminus B_{1/2}.$$

It is easy to verify the existence of a number $N > 0$ so that

$$Q_1 \subset G_N(h, B_1).$$

For a constant ρ_0 to be determined later, we set

$$\vartheta := \rho_0 (u - h).$$

By gathering Propositions 3.14 and 3.12 we obtain $\vartheta \in W^{2,\delta}_{\mathrm{loc}}(B_1)$. Therefore,

$$|A_t(\vartheta, B_1) \cap Q_1| \leq Ct^{-\delta},$$

which follows from the definition of A_t. Since A_N is the complement of G_N, and vice versa,

$$|G_N(u - h, B_1) \cap Q_1| \geq 1 - \rho_0,$$

for some $N > 1$. Finally,

$$|G_{2N}(u, B_1) \cap Q_1| \geq 1 - \rho_0,$$

and the proof is complete.   □

An application of Lemma 3.15 yields the following result.

Lemma 3.16   *Let $u \in C(B_1)$ be a viscosity solutions to (3.4). Suppose the assumptions of Proposition 3.14 are in force and*

$$G_1(u, B_1) \cap Q_3 \neq \emptyset.$$

Then

$$|G_M(u, B_1) \cap Q_1| \geq 1 - \rho,$$

where $M > 0$ and $\rho > 0$ are as in Lemma 3.15.

Compare with the analysis in Proposition 1.44. For a proof of this result, we refer the reader to Caffarelli and Cabré (1995); see also Pimentel and Teixeira (2016, Lemma 5.2).

We continue by introducing the maximal function associated with $f \in L^1_{\mathrm{loc}}(\mathbb{R}^d)$; it is denoted with $m(f)$ and defined as

$$m(f)(x) := \sup_{\ell > 0} \frac{1}{|Q_\ell(x)|} \int_{Q_\ell(x)} |f(y)| \mathrm{d}y.$$

Lemma 3.17   *Let $u \in C(B_1)$ be a viscosity solution to*

$$F_\mu(D^2 u) = f \quad in \quad B_1.$$

Suppose the assumptions of Proposition 3.14 are in force. Extend f outside of
B_1 by zero. Define

$$A := A_{M^{k+1}}(u, B_1) \cap Q_1$$

and

$$B := \left(A_{M^k}(u, B_1) \cap Q_1\right) \cup \left\{x \in Q_1 | m(f^d)(x) \geq (cM^k)^d\right\}.$$

Suppose further there exists $\varepsilon > 0$, yet to be determined, such that

$$\mu + \|f\|_{L^d(B_1)} \leq \varepsilon.$$

Then there exists $\sigma \in (0, 1)$ such that

$$|A| \leq \sigma |B|.$$

As before, for the proof of this result we refer the reader to Caffarelli and
Cabré (1995, Lemma 7.12) and Pimentel and Teixeira (2016, Lemma 5.3). The
integrability of D^2u is closely related to the integrability of Θ, in the sense that

$$\|\Theta\|_{L^p(B_1)} \sim \|D^2u\|_{L^p(B_1)}.$$

See, for instance, Li and Zhang (2015). Therefore, to prove Theorem 3.9, it
suffices to verify that $\Theta \in L^p_{\mathrm{loc}}(B_1)$. Owing to our previous discussion, we are
ultimately interested in the summability of

$$\sum_{k=1}^{\infty} M^{pk} \left|A_{M^k}(u, B_1) \cap Q_1\right|,$$

where $M > 0$ is the constant in Lemma 3.17. We examine this quantity by
resorting to properties of the maximal function associated with $f \in L^p(B_1)$,
combined with the discussion developed so far.

Proof of Theorem 3.9   We take $M > 0$ from Lemma 3.17 and define ρ as
follows:

$$\rho := \frac{1}{2M^p}.$$

In addition, set

$$\alpha_k := \left|A_{M^k}(u, B_1) \cap Q_1\right|$$

and

$$\beta_k := \left|\left\{x \in Q_1 | m(f^d)(x) \geq (CM^k)^d\right\}\right|.$$

Because of Lemma 3.17,

$$\alpha_k \le \rho^k + \sum_{i=0}^{k-1} \rho^{k-i}\beta_i.$$

Moreover, $m(f^d) \in L^{\frac{p}{d}}(\mathbb{R}^d)$ and

$$\left\|m\left(f^d\right)\right\|_{L^{\frac{p}{d}}(\mathbb{R}^d)} \le c\,\|f\|_{L^p(B_1)}^d \le C.$$

Therefore, Proposition 3.13 implies

$$\sum_{k=0}^{\infty} M^{pk}\beta_k \le C.$$

On the other hand we have

$$\mu_\Theta(t) \le \left|A_t(u, B_{1/2})\right| \le \left|A_t(u, B_{1/2}) \cap Q_1\right|.$$

Because of Proposition 3.13, the proof is complete if we verify that

$$\sum_{k=1}^{\infty} M^{pk}\alpha_k \le C.$$

However,

$$\sum_{k=1}^{\infty} M^{pk}\alpha_k \le \sum_{k=1}^{\infty} \left(\rho M^p\right)^k + \sum_{k=1}^{\infty}\sum_{i=0}^{k-1} \rho^{k-i} M^{p(k-i)} M^{pi}\beta_i$$

$$\le \sum_{k=1}^{\infty} 2^{-k} + \left(\sum_{i=0}^{\infty} M^{pi}\beta_i\right)\left(\sum_{j=1}^{\infty} 2^{-j}\right) \le C.$$

□

We continue with a few comments on Theorem 3.9. First, if $f \in L^\infty(B_1)$, it is possible to prove that $D^2u \in \mathrm{BMO}_{\mathrm{loc}}(B_1)$; that is

$$\sup_{\ell>0}\int_{B_\ell} \left|D^2u(x) - \langle D^2u\rangle_\ell\right|^p \mathrm{d}x < \infty,$$

where

$$\langle D^2u\rangle_\ell := \frac{1}{|B_\ell|}\int_{B_\ell} D^2u(x)\mathrm{d}x.$$

Also, Theorem 3.9 extends to operators of the form

$$F : S(d) \times \mathbb{R}^d \times \mathbb{R} \times B_1 \to \mathbb{R},$$

provided F satisfies a structural conditions of the form

$$\mathcal{P}^-_{\lambda,\Lambda}(M-N) - \gamma(x)|p-q| - \omega(|r-s|^+) \leq F(M,p,r,x) - F(N,q,s,x)$$
$$\leq \mathcal{P}^+_{\lambda,\Lambda}(M-N) + \gamma(x)|p-q| + \omega(|r-s|^+),$$

where $M,N \in S(d)$, $p,q \in \mathbb{R}^d$ $r,s \in \mathbb{R}$, $\gamma \in L^\infty(B_1)$, and $\omega\colon \mathbb{R}^+_0 \to \mathbb{R}^+_0$ is a modulus of continuity. In the case where the ingredients in the condition above are continuous, the problem is framed in the context of C-viscosity solutions; conversely, if those ingredients are merely measurable, the L^p-viscosity theory is the correct framework to study the problem. See Caffarelli et al. (1996) and Święch (1997) for details. Finally, we notice that similar arguments produce global estimates, as in Winter (2009), under asymptotic conditions on the problem.

3.2.1 Escauriaza's Exponent

Among the assumptions of Theorem 3.9 is the restriction $p > d$. See Caffarelli (1989) and Pimentel and Teixeira (2016); also, Caffarelli and Cabré (1995, Chapter 7). In 1993, Escauriaza extended Caffarelli's estimates under the condition $p > d - \varepsilon$, for some constant $\varepsilon = \varepsilon(\lambda,\Lambda,d)$.

Proposition 3.18 (Escauriaza's exponent)   *Let $u \in C(B_1)$ be a viscosity solution to (3.4) and suppose that F^* is convex. There exists a universal constant $\varepsilon > 0$ such that, if $p > d - \varepsilon$, then $u \in W^{2,p}_{loc}(B_1)$ and*

$$\|u\|_{W^{2,p}(B_{1/2})} \leq C\left(\|u\|_{L^\infty(B_1)} + \|f\|_{L^p(B_1)}\right).$$

The constant $C > 0$ is universal and $\varepsilon = \varepsilon(\lambda,\Lambda,d)$ is Escauriaza's exponent.

Proposition 3.18 requires lower integrability of the source term to ensure estimates in Sobolev spaces. This weaker requirement is quantified by ε. Although a precise formula for this quantity remains unknown, it depends solely on the dimension and the ellipticity constants. Next, we use the recession strategy to examine some examples of operators and produce asymptotic information on ε.

The key to the lower integrability of the source term is related to F. In fact, it comes from the integrability of the Green function associated with F through its linearized operator L. The following proposition accounts for the integrability of Green's function of a linear (λ,Λ)-elliptic operator. It is due to Fabes and Stroock (1984).

Proposition 3.19 *Let L be a linear (λ, Λ)-elliptic operator with measurable coefficients, and let $G(x, y)$ be its Green function in B_1. Then we have the following.*

(i) *There exists $C > 0$ and $\varepsilon > 0$ such that if $p > d - \varepsilon$,*

$$\int_{B_1} G(x, y)^{p'} \mathrm{d}y \leq C,$$

for all $x \in B_1$, where

$$\frac{1}{p} + \frac{1}{p'} = 1.$$

(ii) *There exists $\beta > 0$ such that if $E \subset B_r \subset B_{1/2}$, we have*

$$\left(\frac{|E|}{|B_r|}\right)^{\beta} \int_{B_r} G(x, y)\mathrm{d}y \leq C \int_E G(x, y)\mathrm{d}y.$$

For the proof of Proposition 3.19 we refer to Fabes and Stroock (1984). This result leads to a Harnack inequality.

Proposition 3.20 (Harnack inequality) *Let $u \in C(B_1)$ be a nonnegative viscosity solution to*

$$F\left(D^2 u\right) = f \quad in \quad B_1,$$

where F is a (λ, Λ)-elliptic operator and $f \in L^{d-\varepsilon}(B_1)$. Then there exists a universal constant $C > 0$ such that

$$\sup_{B_{r/2}} u \leq C \left(\inf_{B_{r/2}} u + r^{2-\frac{d}{d-\varepsilon}} \|f\|_{L^{d-\varepsilon}(B_1)}\right).$$

The proof of Proposition 3.20 is in Escauriaza (1993). This result has many consequences to the general theory of elliptic PDEs. The universal modulus of continuity produced is mentioned in Teixeira (2014b). Indeed, solutions to (3.4) satisfy

$$\|u\|_{C^{0, \frac{d-2\varepsilon}{d-\varepsilon}}(B_{1/2})} \leq C \left(\|u\|_{L^{\infty}(B_1)} + \|f\|_{L^{d-\varepsilon}(B_1)}\right).$$

Notice that Escauriaza's exponent depends only on the integrability of the Green function associated with F and the dimension. Hence, ε is invariant with respect to the ellipticity. In other words, once the dimension d has been fixed, two (λ, Λ)-elliptic operators must have the same exponent ε. Here the recession strategy plays a role. As noted before, in the cases where the limit

$$F^*(M) = \lim_{\mu \downarrow 0} F_{\mu}(M)$$

exists, the recession operator F^* has the same ellipticity as F. If the Green function associated with F^* is known, or we infer its integrability, it would be possible to compute Escauriaza's exponent for F^*, say ε_{F^*}. By knowing this quantity, we recover ε_F. In what follows, we examine an example and explicitly compute Escauriaza's exponent.

Example 3.21 We revisit Example 3.6, where the operator F_q is defined:

$$F_q(M) := \sum_{i=1}^{d} \left(1 + e_i^q\right)^{\frac{1}{q}},$$

for $q \in 2\mathbb{N} + 1$. To linearize this operator and evaluate the integrability of the associated Green function in a ball might be very difficult. However, we learned that $F_q^*(D^2 u) = \Delta u$. In addition, Escauriaza's exponent for the Laplacian operator, denoted with ε_Δ, is known to be $d/2$. Therefore,

$$\varepsilon_{F_q} = \varepsilon_\Delta = \frac{d}{2}.$$

Moreover, we conclude that Theorem 3.9 is available for F_q provided the source term satisfies $f \in L^p(B_1)$, with $p > d/2$.

In the former example, $\varepsilon_{F_q} = d/2$. Every fully nonlinear operator whose recession profile coincides with the Laplacian has the same exponent ε_Δ.

This corpus of results indicates only partially the scope of the asymptotic methods encoded by the recession strategy. Indeed, this approach produces a considerable amount of information also in the context of Hölder spaces. We explore this direction in Section 3.3.

3.3 Applications to Regularity Theory in Hölder Spaces

In the following, we examine the regularity of viscosity solutions to

$$F\left(D^2 u\right) = f \quad \text{in} \quad B_1, \tag{3.6}$$

in Hölder spaces. Here, the operator F is (λ, Λ)-uniformly elliptic and satisfies $F(0) = 0$. Also, the source term is essentially bounded, i.e., $f \in L^\infty(B_1)$. We work under a regularity assumption on the recession function F^*. Namely, we suppose that solutions to the homogeneous equation $F^* = 0$ have $C^{1,1}$-estimates. Under this condition, we prove that solutions to (3.6) are in the function space $C_{\text{loc}}^{1,\,\text{Log-Lip}}(B_1)$, and satisfy

$$\sup_{x \in B_\rho(x_0)} |u(x) - u(x_0) - Du(x_0) \cdot (x - x_0)| \leq -C\rho \ln \rho^2,$$

where $C = C(d, \lambda, \Lambda, \|u\|_{L^\infty(B_1)}, \|f\|_{L^\infty(B_1)})$. To be more precise, we prove the following theorem.

Theorem 3.22 (Regularity in $C^{1,\,\text{Log-Lip}}$) *Let $u \in C(B_1)$ be a viscosity solution to (3.6). Suppose F is a (λ, Λ)-elliptic operator satisfying $F(0) = 0$, Suppose further F^* has $C^{1,1}$-estimates and the limit*

$$\lim_{\mu \to 0} \mu F(\mu^{-1} M) = F^*(M)$$

is uniform in M. Suppose further $f \in L^\infty(B_1)$. Then $u \in C^{1,\,\text{Log-Lip}}_{loc}(B_1)$ and there exists $C > 0$, such that

$$\sup_{x \in B_\rho(x_0)} |u(x) - u(0) - Du(0) \cdot x| \le C\rho^2 \ln \rho^{-1},$$

for every $0 < \rho \ll 1$, with $C = C(d, \lambda, \Lambda, \|u\|_{L^\infty(B_1)}, \|f\|_{L^\infty(B_1)})$.

This result appeared for the first time in Silvestre and Teixeira (2015). Compare with Teixeira (2014b, Theorem 3). The proof of Theorem 3.22 has four parts. First, we verify that a smallness condition on the L^∞-norm of the source term represents no further restriction to the problem. A similar reasoning allows us to consider only normalized solutions to (3.6). The second part of the proof accounts for an oscillation control for a universal radius $0 < r \ll 1$. It is followed by an *iterative argument*. The latter yields an oscillation control at discrete scales of the form $(r^n)_{n \in \mathbb{N}}$. The fourth part puts forward a discrete-to-continuous argument and completes the proof.

As regards the scaling argument, notice that for all $M \in S(d)$, there is $\varepsilon > 0$ such that $\mu < \varepsilon$ implies

$$|F_\mu(M) - F^*(M)| \le \delta,$$

where $\delta > 0$ is chosen as in Proposition 3.23; this is a consequence of Proposition 3.5. We choose $r_0 \sim \sqrt{\varepsilon}$ and define

$$\overline{u}(x) = \frac{\varepsilon u(r_0 x)}{1 + \|u\|_{L^\infty(B_1)} + \|f\|_{L^\infty(B_1)}}.$$

It is clear that $\|\overline{u}\|_{L^\infty(B_1)} \le 1$. In addition,

$$D^2 u(r_0 x) = \frac{\left(1 + \|u\|_{L^\infty(B_1)} + \|f\|_{L^\infty(B_1)}\right) D^2 \overline{u}(x)}{\varepsilon r_0^2}.$$

We then conclude that $\overline{u}$ satisfies

$$\tau F(\tau^{-1} D^2 u_0(x)) = \tau f(r_0 x),$$

with

$$\tau := \frac{\varepsilon r_0^2}{1 + \|u\|_{L^\infty(B_1)} + \|f\|_{L^\infty(B_1)}}.$$

Also,

$$\tilde{f} := \tau f(r_0 x)$$

satisfies

$$\|\tilde{f}\|_{L^\infty(B_1)} \le \varepsilon.$$

Since Theorem 3.22 concerns interior regularity, we state and prove it at the origin. However, a change-of-variables argument of the form $x_0 \mapsto x + x_0$ localizes the analysis in a vicinity of $x_0 \in B_{1/2}$. We start with a proposition controlling the oscillation of solutions.

Proposition 3.23 *Let $u \in C(B_1)$ be a viscosity solution to (3.6). Suppose F is a (λ, Λ)-elliptic operator satisfying $F(0) = 0$, Suppose further F^* has $C^{1,1}$-estimates and the limit*

$$\lim_{\mu \to 0} \mu F(\mu^{-1} M) = F^*(M)$$

is uniform in M. Suppose further $f \in L^\infty(B_1)$. Then there exist a second-order polynomial $P(x)$ and a positive radius $0 < r \ll 1$ satisfying

$$\|u - P\|_{L^\infty(B_r)} \le r^2.$$

Moreover, there exists a universal constant $C > 0$ such that $\|P\|_{L^\infty(B_1)} \le C$.

Proof We use Proposition 3.14 and start by supposing that

$$\mu + \|f\|_{L^\infty(B_1)} \le \varepsilon, \tag{3.7}$$

for some $\varepsilon > 0$ to be determined further in the proof. Let h be the function whose existence follows from Proposition 3.14. Set

$$P(x) := h(0) + Dh(0) \cdot x + x^T \frac{D^2 h(0)}{2} x.$$

Thus

$$\|u - P\|_{L^\infty(B_r)} \le \|u - h\|_{L^\infty(B_r)} + \left\| h - h(0) + Dh(0) \cdot x + x^T \frac{D^2 h(0)}{2} x \right\|_{L^\infty(B_r)}$$

$$\le \delta + C r^{2+\alpha},$$

where $C > 0$ is a universal constant. Now we make universal choices for r and the proximity regime $\delta > 0$ as follows:

$$r := \left(\frac{1}{2C}\right)^{\frac{1}{\alpha}} \qquad \text{and} \qquad \delta := \frac{r^2}{2}.$$

Hence

$$\|u - P\|_{L^\infty(B_r)} \leq r^2,$$

and the proof is complete. The choice of $\delta > 0$ universally determines $\varepsilon > 0$ in (3.7). This constant remains unchanged throughout this section. $\square$

The next proposition extends the oscillation control to discrete scales driven by r.

Proposition 3.24 *Let $u \in C(B_1)$ be a viscosity solution to (3.6). Suppose F is a (λ, Λ)-elliptic operator satisfying $F(0) = 0$, Suppose further that F^* has $C^{1,1}$-estimates and the limit*

$$\lim_{\mu \to 0} \mu F\left(\mu^{-1} M\right) = F^*(M)$$

is uniform in M. Suppose further $f \in L^\infty(B_1)$. Let $r > 0$ be given as in Proposition 3.23. There exists a sequence of quadratic polynomials $(P_n)_{n \in \mathbb{N}}$ of the form

$$P_n(x) := a_n + b_n \cdot x + x^T \frac{M_n}{2} x,$$

where $a_n \in \mathbb{R}$, $b_n \in \mathbb{R}^d$, and $M_n \in S(d)$. In addition, for every $n \in \mathbb{N}$,

$$F^*(M_n) = 0, \tag{3.8}$$

$$\sup_{B_{r^n}} |u - P_n| \leq r^{2n} \tag{3.9}$$

and

$$|a_n - a_{n-1}| + r^{n-1} |b_n - b_{n-1}| + r^{2(n-1)} |M_n - M_{n-1}| \leq C r^{2(n-1)}, \tag{3.10}$$

for a universal constant $C > 0$.

Proof We use an induction argument and split the proof into three steps.

Step 1. We start with the base case, $n = 1$. Set $P_0 \equiv 0$ and

$$P_1(x) := h(0) + Dh(0) \cdot x + x^T \frac{D^2 h(0)}{2} x,$$

where h is the approximating function whose existence follows from Proposition 3.14. Proposition 3.23 is precisely the case $n = 1$. Suppose we have established the case $n = k$; next we examine the case $n = k + 1$.

Step 2. We introduce the auxiliary function

$$v_k(x) := \frac{u(r^k x) - P_k(r^k x)}{r^{2k}}.$$

The induction hypothesis ensures $\|v\|_{L^\infty(B_1)} \leq 1$. Also, we have

$$D^2 v(x) = D^2 u(r^k x) - M_k.$$

As a consequence, v_k is a viscosity solution to

$$\mu F\left(\mu^{-1}\left(D^2 v + M_k\right)\right) = f(r^k x).$$

Let

$$F_k(M) := F(M + M_k)$$

and

$$F_k^*(M) := F^*(M + M_k).$$

Define also

$$F_{\mu,k}(M) := \mu F\left(\mu^{-1} M + M_k\right).$$

it follows that

$$\left| F_{\mu,k}(M) - F_k^*(M) \right| \leq \delta.$$

Because $F^*(M_k) = 0$, the equation $F_k^* = 0$ has the same $C^{1,1}$-estimates as $F^* = 0$. Therefore, there exists a $C^{1,1}$-function $\tilde{h}$, with the same bounds as h, δ-approximating v_k. Proposition 3.23 implies the existence of $\tilde{P}(x)$, given by

$$\tilde{P}(x) := \tilde{h}(0) + D\tilde{h}(0) \cdot x + x^T \frac{D^2 \tilde{h}(0)}{2} x,$$

with

$$\left| \tilde{h}(0) \right| + \left| D\tilde{h}(0) \right| + \left| D^2 \tilde{h}(0) \right| \leq C, \tag{3.11}$$

satisfying

$$\left\| v_k - \tilde{P} \right\|_{L^\infty(B_r)} \leq r^2.$$

Observe that $C > 0$ in (3.11) is the same constant as in Proposition 3.23. The definition of v_k implies

$$\sup_{x \in B_r} \frac{\left| u\left(r^k x\right) - P_k\left(r^k x\right) - r^{2k} \tilde{P}(x) \right|}{r^{2k}} \le r^2;$$

that is,

$$\sup_{x \in B_{r,k+1}} \left| u(x) - \left(P_k(x) + r^{2k} \tilde{P}\left(r^{-1}x\right)\right) \right| \le r^{2(k+1)}.$$

We proceed by defining P_{k+1} as

$$P_{k+1}(x) := P_k(x) + r^{2k} \tilde{P}\left(r^{-k}x\right),$$

which leads to (3.9) in the case $n = k + 1$. We still have to ensure that (3.8) and (3.10) hold for $n = k + 1$.

Step 3. The definition of $P_{k+1}(x)$ yields

$$a_{k+1} - a_k = r^{2k}\tilde{h}(0), \qquad \cdot$$

$$r^k \left| b_{k+1} - b_k \right| = r^{2k} \left| D\tilde{h}(0) \right|,$$

and

$$r^{2k} \left\| M_{k+1} - M_k \right\| = r^{2k} \left\| D^2\tilde{h}(0) \right\|.$$

Combining the former with (3.11) we verify (3.10) for $n = k + 1$. To conclude that $F^*(M_{k+1}) = 0$ we notice that

$$F^*\left(M_{k+1}\right) = F^*\left(D^2\tilde{h}(0) + M_k\right) = F_k^*\left(D^2\tilde{h}(0)\right) = 0,$$

which completes the proof. $\qquad\qquad\qquad\qquad\qquad\qquad\qquad\qquad\square$

To prove Theorem 3.22 we switch from an oscillation control at discrete scales to the continuous setting.

Proof of Theorem 3.22   For $0 < \rho \ll 1$, let $n \in \mathbb{N}$ be such that $r^{n+1} < \rho \leq r^n$. Hence,

$$\sup_{B_\rho} |u(x) - (u(0) + Du(0) \cdot x)| \leq \sup_{B_{r^n}} |\theta(x) - (u(0) + Du(0) \cdot x)|$$

$$\leq \sup_{B_{r^n}} \left| (u - P_n) + a_n - u(0) + b_n \cdot x - Du(0) \cdot x + \frac{x^T M_n x}{2} \right|$$

$$\leq r^{2n} + r^{2n} + C r^{2n} + \frac{C}{2} n r^{2n}$$

$$\leq C \left(r^{2n} + n r^{2n} \right)$$

$$\leq \frac{C}{r^2} \left(r^{2(n+1)} + n r^{2(n+1)} \right)$$

$$\leq \frac{C}{r^2} \left(\rho^2 + \frac{\ln \rho}{\ln r} \rho^2 \right)$$

$$\leq \frac{C}{r^2 |\ln \rho|} \left(|\ln \rho| + \ln \frac{1}{r} \right) r^2$$

$$\leq \frac{2C}{r^2 |\ln r|} r^2 \ln \frac{1}{\rho}$$

$$\leq C \rho^2 \ln \frac{1}{\rho},$$

which finishes the poof.   □

3.4 Weak Regularity Theory: Density Results

In what follows we prove a weak regularity result. We refer to the density of more regular solutions in the class of viscosity solutions as *weak regularity*.

The optimal regularity associated with $F(D^2 u) = f$ is the Hölder continuity of Du. Meanwhile, in some cases solutions can be approximated by more regular functions. For example, it is known that $W^{2,p}_{\text{loc}}(B_1) \cap S(\lambda^-, \Lambda^+, f)$ is dense in the class of C-viscosity solutions $S(\lambda, \Lambda, f)$; we remark that this is not known in the class of L^p-viscosity solutions. Therefore, when studying properties closed under uniform limits, the starting point of the theory shifts to $W^{2,p}$-estimates (Pimentel and Teixeira, 2016).

This class of results benefits from the recession operator associated with a sequence of operators. In fact, given $F \colon S(d) \to \mathbb{R}$, we can affect its recession profile by changing it outside of a large ball $B_R \subset S(d)$, for $R \gg 1$. Next, we explore this idea and detail a weak regularity result concerning $C^{1, \text{Log-Lip}}$-estimates.

Theorem 3.25 (Weak estimates in $C^{1, \text{Log-Lip}}$) *Let* $u \in C(B_1)$ *be a C-viscosity solution to*

$$F(D^2 u) = f \qquad in \qquad B_1. \tag{3.12}$$

Suppose F is a (λ, Λ)-elliptic operator $F(0) = 0$, and $f \in C(B_1) \cap L^\infty(B_1)$. Then there exists a sequence of functions $(u_n)_{n \in \mathbb{N}} \subset C(B_1)$ such that

$$(u_n)_{n \in \mathbb{N}} \subset C^{1, \text{Log-Lip}}_{loc}(B_1) \cap S(\lambda^-, \Lambda^+, f)$$

and $u_n \to u$ locally uniformly in B_1.

Proof We construct a sequence of operators and examine their recession functions. We split the proof into three steps.

Step 1. Fix $\delta > 0$ and let $F_n = F_n(M)$ be defined as

$$F_n(M) := \max \left\{ F(M), \mathcal{P}^+_{\lambda - \delta, \Lambda + \delta}(M) - C_n \right\},$$

where $(C_n)_{n \in \mathbb{N}}$ is a sequence of positive numbers determined further. For simplicity, denote $\mathcal{P}^+_\delta := \mathcal{P}^+_{\lambda - \delta, \Lambda + \delta}$. We compute a lower bound for $F(M)$ as follows:

$$F(M) \geq -\Lambda \operatorname{Tr}(M^+) + \lambda \operatorname{Tr}(M^-) + \mathcal{P}^+_\delta(M) - \mathcal{P}^+_\delta(M)$$
$$\geq \mathcal{P}^+_\delta(M) - (\Lambda - \lambda + \delta) \|M\|,$$

where

$$\|M\| := \operatorname{Tr}(M^+) + \operatorname{Tr}(M^-).$$

Set $C_n := n(\Lambda - \lambda + \delta)$; for $\|M\| \leq n$ we have

$$F(M) \geq \mathcal{P}^+_\delta(M) - C_n,$$

Therefore, the operator F_n coincides with F in $B_n \subset S(d)$, for every $n \in \mathbb{N}$: i.e.,

$$F_n = F \qquad in \qquad B_n \subset S(d).$$

As a consequence, $F_n \to F$ uniformly as $n \to \infty$. Now we turn our attention to the behavior of F_n at the ends of $S(d)$.

Step 2. Consider

$$F_{n, \mu}(M) = \mu F_n(\mu^{-1} M) = \max \left\{ F_\mu(M), \mathcal{P}^+_\delta(M) - \mu C_n \right\}.$$

The (λ, Λ)-ellipticity of F_μ yields

$$F_\mu(M) \leq \mathcal{P}^+_{\lambda, \Lambda}(M) + \mathcal{P}^+_\delta(M) - \mathcal{P}^+_\delta(M)$$
$$\leq \mathcal{P}^+_\delta(M) - \delta \|M\|.$$

By taking

$$\|M\| \geq \frac{\mu C_n}{\delta},$$

we conclude

$$F_\mu(M) \leq L_\delta(M) - \mu C_n,$$

Therefore,

$$F_n^* = \lim_{\mu \to 0} F_{n,\mu} = \lim_{\mu \to 0} (L_\delta(M) - \mu C_n) = L_\delta(M).$$

We conclude that F_n has a recession profile for which $C^{1,1}$-estimates are available. In the following we combine the previous steps and Theorem 3.22.

Step 3. Let $(u_n)_{n \in \mathbb{N}} \subset C(B_1)$ be a sequence of viscosity solutions to

$$F_n(D^2 u_n) = f \qquad \text{in} \qquad B_1. \tag{3.13}$$

From Theorem 3.22, we learn that $u_n \in C_{\mathrm{loc}}^{1, \, \mathrm{Log\text{-}Lip}}(B_1)$ and there exists $C_n > 0$ such that

$$\sup_{x \in B_\rho} |u_n(x) - (u_n(0) + Du_n(0) \cdot x)| \leq C_n \rho^2 \ln \frac{1}{\rho},$$

for every $0 < \rho \ll 1$. Finally, equip (3.13) with Dirichlet boundary conditions of the form

$$u_n = u \qquad \text{on} \qquad \partial B_1.$$

Although the bounds for $(u_n)_{n \in \mathbb{N}}$ in $C_{\mathrm{loc}}^{1, \, \mathrm{Log\text{-}Lip}}(B_1)$ depend on n, the sequence is equibounded in $C^{1,\alpha}(B_1)$, for some $\alpha \in (0,1)$; see Theorem 1.56. Hence, it admits a subsequence, still denoted with $(u_n)_{n \in \mathbb{N}}$, converging locally uniformly to some $u_\infty \in C_{\mathrm{loc}}^{1, 1/2}(B_1)$. In addition, $F_n \to F$ locally uniformly in $S(d)$. Theorem 1.17 implies $u_\infty \equiv u$; the uniqueness of C-viscosity solutions completes the proof. $\square$

Remark 3.26   The uniqueness of C-viscosity solutions is critical in the previous proof. We complete this chapter with a discussion on the limitations of the approach based on the recession operator.

3.5 Limitations of the Recession Strategy

The approach based on the recession operator produces important regularity results. In the context of approximation methods, F^* plays the role of the fixed-coefficients operator $F(x_0, M)$ in Caffarelli (1989); c.f. Silvestre and Teixeira (2015). However, the recession approach also faces important limitations. In the following we discuss three instances where the recession operator falls short.

Let $F: S(d) \rightarrow \mathbb{R}$ be a (λ, Λ)-elliptic operator satisfying a homogeneity condition of the form

$$F(tM) = tF(M),$$

for every $t \geq 0$. It is clear that

$$\mu F(\mu^{-1}M) = \mu\mu^{-1}F(M) = F(M).$$

That is, in the case of positively homogeneous operators, the recession strategy provides no further information on the problem, as $F_\mu \equiv F$ for every $\mu > 0$. This observation affects important classes of problems in the theory. We highlight the case of the Isaacs equation.

Example 3.27 (The Isaacs equation)   The Isaacs equation is an important fully nonlinear elliptic model. It can be written as

$$F(M) := \sup_{\alpha \in \mathcal{A}} \inf_{\beta \in \mathcal{B}} \left[-\text{Tr}(A_{\alpha,\beta}(x)M) \right] = 0,$$

where $A_{\alpha,\beta}: B_1 \rightarrow S(d)$ is a matrix-valued function, depending on parameters $\alpha \in \mathcal{A}$ and $\beta \in \mathcal{B}$ and satisfying a (λ, Λ)-ellipticity condition. For simplicity, we suppose the parameter spaces $\mathcal{A}$ and $\mathcal{B}$ are compact metric spaces.

This formulation appears in stochastic optimal control; more precisely, in the analysis of two-players, zero-sum stochastic differential games. The Isaacs operator also plays an important role in the theory of fully nonlinear problems. The Isaacs operator is homogeneous of degree 1. Therefore,

$$\mu \sup_{\alpha \in \mathcal{A}} \inf_{\beta \in \mathcal{B}} \left[-\text{Tr}(A_{\alpha,\beta}(x)\mu^{-1}M) \right] = \sup_{\alpha \in \mathcal{A}} \inf_{\beta \in \mathcal{B}} \left[-\text{Tr}(A_{\alpha,\beta}(x)M) \right]$$

and

$$F \equiv F_\mu \equiv F^*;$$

i.e., the recession strategy produces no further information in this case.

A second limitation for the recession operator is the level of regularity it produces. Contrary to the fixed-coefficients approach (Caffarelli, 1989), it is not possible to prove a C^2-regularity result through assumptions on F^*. To verify this fact, we recall one of the counterexamples discussed in Section 1.7.

In fact, there exists $v \in C^{1,1}_{loc}(B_1) \setminus C^2_{loc}(B_1)$ and a (λ, Λ)-elliptic operator F such that

$$F(D^2v) = 0 \quad \text{in} \quad B_1.$$

Let

$$R := \left(\frac{\Lambda + \lambda}{\lambda}\right) \|D^2v\|_{L^\infty(B_1)}.$$

Define $\overline{F}$ as the (λ, Λ)-elliptic operator that coincides with F in $B_R \subset S(d)$ but is modified outside of $B_{100R} \subset S(d)$ such that $\overline{F}^*(M) \equiv \text{Tr}(M)$. We claim that v is a viscosity solution to

$$\overline{F}(D^2v) = 0 \quad \text{in} \quad B_1.$$

If we verify the claim, v would be a nonclassical solution to $\overline{F} = 0$, even under the condition that $\overline{F}^*$ is the Laplacian operator. We proceed by verifying the claim in the case of subsolutions. Suppose by contradiction that v is not a subsolution to $\overline{F} = 0$. Then, there exists $x_0 \in B_1$ and $\varphi \in C^2(B_1)$ such that $v - \varphi$ attains a local maximum at x_0, but

$$\overline{F}(D^2\varphi(x_0)) > 0.$$

Suppose φ is of the form

$$\varphi(x) := \ell(x) + x^T \frac{M}{2} x,$$

where $\ell(x)$ is an affine function and M is a symmetric matrix. In this case, the contradiction hypothesis reads

$$\overline{F}(M) > 0. \tag{3.14}$$

If $\|M\| \leq R$, $\overline{F}(M) = F(M)$ and (3.14) is false, because v solves $F = 0$. Hence, (3.14) implies $\|M\| > R$.

The Bony maximum principle, as stated in Lions (1983), yields

$$\text{ess} \lim_{x \to x_0} \inf e_i\left(D^2v(x)\right) \leq e_i(M),$$

where $e_i(N)$ denotes the ith eigenvalue of the matrix N. Also, (3.14) implies that M has at least one negative eigenvalue. Ellipticity then leads to

$$0 < -\lambda \text{Tr}\left(M^+\right) + \Lambda \text{Tr}\left(M^-\right) \leq -\lambda \text{Tr}\left(M^+\right) + \Lambda \sup_{x \in B_1} \|D^2v(x)\|.$$

Hence,

$$\|M\| \leq \frac{\Lambda}{\lambda} \|D^2v(x_0)\| + \|D^2v(x_0)\| = R,$$

which is a contradiction with $\|M\| > R$ and yields the claim. We conclude by discussing the use of the recession strategy in the context of improved regularity for flat solutions.

An intermediate step to the partial regularity result is the $C^{2,\alpha}$-regularity of flat solutions to $F = 0$. Armstrong et al. (2012) prove the following result.

Theorem 3.28 (Flat solutions are $C^{2,\alpha}$-regular)   *Let $u \in C(B_1)$ be a viscosity solution*

$$F(D^2 u) = 0 \quad in \quad B_1.$$

Suppose F is (λ, Λ)-elliptic; suppose further $F \in C^1(S(d))$ and DF has a modulus of continuity denoted with ω_F. Given $\alpha \in (0,1)$, there exists $0 < \sigma \ll 1$ such that, if $\|u\|_{L^\infty(B_1)} \leq \sigma$, then $u \in C^{2,\alpha}_{loc}(B_1)$. In addition,

$$\|u\|_{C^{2,\alpha}(B_{1/2})} \leq C \|u\|_{L^\infty(B_1)},$$

where $C > 0$ is universal and $\sigma = \sigma(d, \lambda, \Lambda, \alpha, \omega_F)$.

As we have discussed, one usually displaces the assumption from F to its the recession function F^* in an attempt to produce regularity in a more general framework. This strategy falls short in the context of Theorem 3.28. In fact, one should suppose the differentiability condition required in the theorem is met by F^*. However, in case F^* is unique, it is homoegeneous of degree one. Being also differentiable, it would be linear. Hence, the flatness regime at the level of F^* becomes redundant.

An alternative would be to drop the uniqueness of F^* and suppose that every subssequential limit of F_μ, as $\mu \to 0$, meets the same differentiability condition. Under these conditions, every subsequential limit would be linear. Indeed, if F^* is a subsequential limit of F_μ, so is F^*_μ. Moreover,

$$DF^*_\mu(M) = DF^*(\mu^{-1}M).$$

For DF^* and DF^*_μ to have the same modulus of continuity, DF^*_μ must be independent of μ.

Bibliographical Notes

The notion of recession function appears in the context of convex analysis; we refer the reader to the book by Rockafellar (1997). For instance, if the recession profile associated with a finite convex function $g: \mathbb{R}^d \to \mathbb{R}$ is finite, then g is Lipschitz-continuous (Rockafellar, 1997, Theorem 10.5). The use of the recession strategy in regularity theory was introduced in the work of Silvestre and Teixeira (2015), where the authors develop the Hölder regularity theory, paralleling Caffarelli's program launched in 1989 (see Caffarelli, 1989). A Sobolev regularity theory through the recession technology is developed in Pimentel and Teixeira (2016). This effort also relates to Caffarelli (1989) as it replaces the fixed-coefficients counterpart of the original operator with the recession profile. Pimentel and Teixeira (2016) prove a *weak* regularity result, in the spirit of Theorem 3.25, in the context of Sobolev spaces. This type of result is very much inspired by the analysis due to Caffarelli and Silvestre (2010b). In Sections 3.2 and 3.3 we have addressed regularity in Sobolev/BMO and Hölder spaces, respectively. Recently, we learned from Mark Allen about a connection between functions with derivatives in BMO and the Zygmund class. In fact, a function with a derivative in BMO belongs to the Zygmund class; as a consequence, it has a Log-Lip-modulus of continuity; see Nicolau and Soler i Gibert (2019) and Zygmund (2002). Hence, once the results in Section 3.2 have been obtained, we infer the findings put forward in Section 3.3 from this observation.

4

A Regularity Theory for the Isaacs Equation

In this chapter we present a regularity theory for the solutions of the Isaacs equation through an approximation strategy. By imposing conditions on the coefficients of the problem, we approximate the Isaacs operator with a Bellman one, with fixed coefficients. Different proximity assumptions lead to distinct regularity regimes, covering estimates in Hölder and Sobolev spaces.

We examine L^p-viscosity solutions to an Isaacs equation of the form

$$\sup_{\alpha \in \mathcal{A}} \inf_{\beta \in \mathcal{B}} \left(- \operatorname{Tr} \left(A_{\alpha, \beta}(x) D^2 u \right) \right) = f \quad \text{in} \quad \Omega, \tag{4.1}$$

where $A_{\alpha, \beta} \colon B_1 \to S(d)$ is a (λ, Λ)-elliptic matrix for every $\alpha \in \mathcal{A}$ and $\beta \in \mathcal{B}$, $f \in L^\infty(\Omega)$, and $\mathcal{A}$ and $\mathcal{B}$ are compact, separable, and countable metric spaces.

Our strategy to produce a regularity theory for the solutions to (4.1) is to impose a proximity regime on the coefficients $A_{\alpha, \beta}(\cdot)$. To be precise, we suppose they are close to a constant matrix A_β; this choice relates (4.1) to a Bellman operator, which is known to be convex.

4.1 Some Context

The Isaacs equation appeared for the first time in the works of R. Isaacs, in close connection with optimal control problems; in particular, in the context of two-player, zero-sum differential games (Isaacs, 1965; Friedman, 1971). Modeled at first in the deterministic setting, those games give rise to a first-order Isaacs equation; its second-order elliptic counterpart arises in the stochastic framework. In fact, the operator

$$- \operatorname{Tr}(A_{\alpha, \beta}(x) M)$$

stands for the infinitesimal generator of the stochastic dynamics governing the state of the game. The remarkable aspect is that solutions to the Isaacs equation are the value function of the associated differential game. This connection can be made rigorous through an application of the dynamic programming principle, together with further considerations from game theory. One finds applications of this class of games in various disciplines.

An important question arising in the analysis of (4.1) comes precisely from the structure of the two-player zero-sum games. Roughly speaking, two players face a payoff functional and choose controls to optimize it. Player I chooses $\alpha \in \mathcal{A}$ aiming at maximizing the payoff functional. Simultaneously, Player II chooses $\beta \in \mathcal{B}$ in order to minimize it. An important step in the mathematical treatment of these models is in the understanding of the term *simultaneously*.

In fact, one can suppose that, at an initial instant t_0, Player I chooses a control α to be in force in the time interval $[t_0, t_0 + h =: t_1)$, for some $h > 0$. Then Player II chooses β for the interval $[t_1, t_1 + h =: t_2)$, and Player I makes a new choice having in mind the time interval $[t_2, t_2 + h)$. The process repeats until it exhausts the time span of the game. Here, Player II acts with knowledge of Player I's choices. Similarly, one can reverse the previous scheme and suppose Player II is the one choosing controls $\beta \in \mathcal{B}$ first. As $h \to 0$, we recover a scenario where players make decisions simultaneously.

Depending on *which player acts first*, the Isaacs equation changes. If Player I is the first to play, it becomes

$$\inf_{\beta \in \mathcal{B}} \sup_{\alpha \in \mathcal{A}} \left(- \operatorname{Tr} \left(A_{\alpha, \beta}(x) D^2 u \right) \right) = f \quad \text{in} \quad \Omega;$$

otherwise, it is given by

$$\sup_{\alpha \in \mathcal{A}} \inf_{\beta \in \mathcal{B}} \left(- \operatorname{Tr} \left(A_{\alpha, \beta}(x) D^2 u \right) \right) = f \quad \text{in} \quad \Omega.$$

A solution $\underline{v}$ to the former is called lower value of the game, whereas a solution $\overline{v}$ to the latter is the upper value. An important property regards the coincidence of both quantities, i.e., $\underline{v} = \overline{v}$. In this case, the game is said to have a value. It is tantamount to saying that the outcome of the game does not depend on the order in which players make decisions.

The conditions for this equality to hold neatly bridges the realm of Isaacs equations and the theory of viscosity solutions. Indeed, if the so-called Isaacs condition holds, that is,

$$\sup_{\alpha \in \mathcal{A}} \inf_{\beta \in \mathcal{B}} \left(- \operatorname{Tr} \left(A_{\alpha, \beta}(x) M \right) \right) = \inf_{\beta \in \mathcal{B}} \sup_{\alpha \in \mathcal{A}} \left(- \operatorname{Tr} \left(A_{\alpha, \beta}(x) M \right) \right).$$

for every $M \in S(d)$, the uniqueness of viscosity solutions implies that $\overline{v} = \underline{v}$ and the game has a value (Barron et al., 1984). We notice the theory of viscosity solutions led to important advances in the understanding of (4.1).

In addition to its connection with game theory and viscosity solutions, we mention another feature of the Isaacs operator. Let $F\colon S(d) \to \mathbb{R}$ be a (λ, Λ)-elliptic operator. As pointed out by Evans (2007b), $F(M)$ can be written as

$$F(M)\colon \min_{\beta \in \mathcal{B}} c_\beta(M),$$

where $c_\beta\colon S(d) \to \mathbb{R}$ are convex functions and $\mathcal{B}$ is a set. At this point, we notice that every convex function c_β can be written as

$$c_\beta(M) = \max_{\alpha \in \mathcal{A}} \left(a_{\alpha, \beta} \cdot M + b_{\alpha, \beta} \right),$$

where $a_{\alpha, \beta} \in \mathbb{R}^{\frac{d(d+1)}{2}}$, $b_{\alpha, \beta} \in \mathbb{R}$, and $\mathcal{A}$ is a set. As a consequence, any fully nonlinear elliptic operator can be written as

$$F(M) = \min_{\beta \in \mathcal{B}} \max_{\alpha \in \mathcal{A}} \left(a_{\alpha, \beta} \cdot M + b_{\alpha, \beta} \right),$$

which is a form of Isaacs operator. See also Cabré and Caffarelli (2003, Remark 1.5).

The motivation for the study of a regularity theory for the L^p-viscosity solutions to (4.1) sits in various aspects. First, the operator driving the equation lacks convexity/concavity, since it combines the infimum and the supremum. As a consequence, the problem falls out of the scope of the Evans–Krylov regularity theory. Moreover, the operator fails differentiability and the partial regularity result is not available to (4.1). Finally, it is known that there exists a (λ, Λ)-elliptic operator of Isaacs type whose regularity theory is strictly below $C^{1,1}$. See Section 1.7; see also Nadirashvili et al. (2014, Chapter 4.4). Therefore, a natural problem is to understand conditions unlocking improved regularity results for (4.1).

Standard strategies fall short in the context of the Isaacs equation. As mentioned before, the recession operator is redundant in the context of the Isaacs equation, since the operator is positively homogeneous of degree one.

Another technique to study the regularity of (4.1) is to consider its fixed-coefficients counterpart and develop the program in Caffarelli (1989) and Caffarelli and Cabré (1995). Meanwhile, the fixed-coefficients operator becomes

$$\sup_{\alpha \in \mathcal{A}} \inf_{\beta \in \mathcal{B}} \left(-\operatorname{Tr} \left(A_{\alpha, \beta}(x_0) M \right) \right).$$

The operator also lacks concavity/convexity; it may even have singular solutions. As a consequence, it is unreasonable to expect $C^{1,1}$ or even $W^{2,q}$-regularity estimates for it.

To bypass these genuine difficulties, we approximate the Isaacs equation by a Bellman equation

$$\inf_{\beta \in \mathcal{B}} \left(-\operatorname{Tr}\left(A_\beta(x)D^2 u\right)\right) = 0 \quad \text{in} \quad \Omega,$$

where $A_\beta(\cdot)$ is a (λ, Λ)-elliptic matrix. For simplicity we may assume $A_\beta(x) \equiv A_\beta$ to be constant. Notice that in this case, the Bellman operator is concave; hence, its homogeneous equation has $C^{2,\alpha}$-estimates, as implied by the Evans–Krylov theory. Moreover, suppose a given Isaacs operator has a *very small* dependence on one of the controls – say,

$$\sup_{\alpha \in \mathcal{A}} A_{\alpha,\beta} - \inf_{\alpha \in \mathcal{A}} A_{\alpha,\beta} \ll 1,$$

uniformly in $\beta \in \mathcal{B}$. From a heuristic viewpoint, such a condition suggests the Isaacs equation associated with $A_{\alpha,\beta}$ is *almost* a Bellman equation.

Bearing this idea in mind we aim at finding conditions on the coefficients $A_{\alpha,\beta}(\cdot)$ governing (4.1) to transmit regularity from the Bellman operator through an approximation strategy. In Section 4.2 we discuss aspects of the Bellman equation useful to our analysis.

4.2 The Bellman Equation

Given a countable set $\mathcal{B}$, we consider a family of uniformly elliptic matrix-valued functions $(A_\beta(\cdot))_{\beta \in \mathcal{B}}$ and a family of vector fields $(b_\beta)_{\beta \in \mathcal{B}}$. More precisely, for $\beta \in \mathcal{B}$, let $A_\beta \colon \Omega \to S(d)$ be such that

$$\lambda I \le A_\beta(x) \le \Lambda I, \tag{4.2}$$

for every $(\beta, x) \in \mathcal{B} \times \Omega$. Also, suppose $b_\beta \colon \Omega \to \mathbb{R}^d$ satisfies

$$\sup_{\beta \in \mathcal{B}} |b_\beta(x)| \le \gamma, \tag{4.3}$$

uniformly in $x \in \Omega$, for some $\gamma > 0$ fixed. The Bellman equation associated with A_β and b_β is

$$\inf_{\beta \in \mathcal{B}} \left(-\operatorname{Tr}\left(A_\beta(x)D^2 u\right) + b_\beta(x) \cdot Du \right) = f \quad \text{in} \quad \Omega, \tag{4.4}$$

where $f \in L^p(\Omega)$, for some $p > d/2$.

Under (4.2) and (4.3), the Bellman equation satisfies the structure condition in Definition 1.12. In fact, for $M, N \in S(d)$, and $p, q \in \mathbb{R}^d$ we have

$$\inf_{\beta \in \mathcal{B}} \left(-\operatorname{Tr}\left(A_\beta(x)M\right) + b_\beta(x) \cdot p\right) - \inf_{\beta \in \mathcal{B}} \left(-\operatorname{Tr}\left(A_\beta(x)N\right) + b_\beta(x) \cdot q\right)$$

$$\leq \inf_{\beta \in \mathcal{B}} \left(-\operatorname{Tr}\left(A_\beta(x)M\right)\right) - \inf_{\beta \in \mathcal{B}} \left(-\operatorname{Tr}\left(A_\beta(x)N\right)\right) + \gamma |p - q|$$

$$\leq \sup_{\lambda I \leq A \leq \Lambda I} \left(-\operatorname{Tr}(A(M - N))\right) + \gamma |p - q|$$

$$= \mathcal{P}^+_{\lambda, \Lambda}(M - N) + \gamma |p - q|.$$

The remaining inequality follows from a similar reasoning. For simplicity, we suppose $A_\beta(x) \equiv A_\beta$ is a constant matrix and $b_\beta \equiv 0$. In this case, we notice the Bellman equation is governed by a concave operator.

As a consequence, it falls within the scope of the Evans–Krylov regularity theory; see Section 1.5. Hence, classical solutions are available to the problem. More precisely, we have the following corollary.

Corollary 4.1 ($C^{2,\alpha}$-regularity for the Bellman equation) *Let $u \in C(\Omega)$ be an L^p-viscosity solution to*

$$\inf_{\beta \in \mathcal{B}} \left(-\operatorname{Tr}\left(A_\beta D^2 u\right)\right) = 0 \quad in \quad \Omega,$$

where A_β satisfies (4.2). Then $u \in C^{2,\alpha}_{loc}(\Omega)$ and there exists $C > 0$ such that, for every $\Omega' \Subset \Omega$,

$$\|u\|_{C^{2,\alpha}(\Omega')} \leq C \|u\|_{L^\infty(\Omega)}.$$

The constant $C > 0$ depends on $d, \lambda, \Lambda, \operatorname{diam}(\Omega)$, and $\operatorname{dist}(\Omega', \partial\Omega)$.

In what follows we prove regularity results for an Isaacs equation of the form

$$\sup_{\alpha \in \mathcal{A}} \inf_{\beta \in \mathcal{B}} \left(-\operatorname{Tr}\left(A_{\alpha,\beta}(x)D^2 u\right) + b_{\alpha,\beta}(x) \cdot Du\right) = f \quad in \quad \Omega$$

by assuming there exists a matrix A_β such that

$$\left|A_{\alpha,\beta}(x) - A_\beta\right|$$

is controlled in some suitable sense (to be made precise further) and that $\|b_{\alpha,\beta}\|_{L^\infty(\Omega)} \leq C$, for some $C > 0$.

As a result, we transmit information from

$$\inf_{\beta \in \mathcal{B}} \left(-\operatorname{Tr}\left(A_\beta D^2 u\right)\right) = 0$$

to our original problem. The relevant information in this setting is precisely the statements in Corollary 4.1. We start with regularity estimates in Sobolev spaces.

4.3 Regularity for the Isaacs Equation in Sobolev Spaces

To prove a regularity result in Sobolev spaces $W^{2,p}_{\mathrm{loc}}(\Omega)$ for L^p-viscosity solutions to

$$\sup_{\alpha\in\mathcal{A}}\inf_{\beta\in\mathcal{B}}\left(-\operatorname{Tr}\left(A_{\alpha,\beta}(x)D^2u\right)-b_{\alpha,\beta}(x)\cdot Du\right)=f \quad \text{in}\quad \Omega, \quad (4.5)$$

we impose a few conditions on the data of the problem.

Assumption 4.2 (Coefficients $A_{\alpha,\beta}$)   There exist constants $0<\lambda\le\Lambda$ such that

$$\lambda I \le A_{\alpha,\beta}(x) \le \Lambda I,$$

for every $\alpha\in\mathcal{A}$, $\beta\in\mathcal{B}$, and $x\in\Omega$. In addition, $A_{\alpha,\beta}(\cdot)$ is uniformly continuous, in the parameters $\alpha\in\mathcal{A}$ and $\beta\in\mathcal{B}$.

We proceed with an assumption on the proximity of $A_{\alpha,\beta}$ with respect to a family of matrices depending only on $\beta\in\mathcal{B}$.

Assumption 4.3 (Bellman coefficients I)   We suppose there exists a family $(A_\beta)_{\beta\in\mathcal{B}}$ such that

$$\lambda I \le A_\beta \le \Lambda I,$$

for every $\beta\in\mathcal{B}$. In addition, there exists $0<\varepsilon_1\ll1$, small enough, such that

$$|A_{\alpha,\beta}(x)-A_\beta|\le\varepsilon_1,$$

uniformly in x, α, and β. The parameter ε_1 is determined in Lemma 4.12; see also Remark 4.13.

Our next assumption concerns the integrability of the family $(b_{\alpha,\beta})_{\alpha\in\mathcal{A},\beta\in\mathcal{B}}$. It reads as follows.

Assumption 4.4 (Vector field $b_{\alpha,\beta}$)   We suppose the family $(b_{\alpha,\beta})_{\alpha\in\mathcal{A},\beta\in\mathcal{B}}\subset L^\infty(\Omega,\mathbb{R}^d)$, and there exists $C>0$ such that

$$\|b_{\alpha,\beta}\|_{L^\infty(\Omega,\mathbb{R}^d)}\le C,$$

for every $\alpha\in\mathcal{A}$ and $\beta\in\mathcal{B}$.

Because we relate our operator to an homogeneous Bellman equation, we require some integrability for the source term f.

Assumption 4.5 (Integrability of the source term)   We suppose $f \in L^p(\Omega)$, where $p > d$.

Under these conditions we prove that L^p-viscosity solutions to (4.5) are in $W^{2,p}_{\text{loc}}(\Omega)$ and the appropriate estimates are available. This is the content of the next theorem.

Theorem 4.6 (Estimates in Sobolev spaces)   *Let $u \in C(\Omega)$ be an L^p-viscosity solution to (4.5), where $p > d$ is fixed. Suppose Assumptions 4.2–4.5 hold true. Then, $u \in W^{2,p}_{\text{loc}}(\Omega)$ and, for every $\Omega' \Subset \Omega$ there exists $C > 0$ such that*

$$\|u\|_{W^{2,p}(\Omega')} \leq C \left(\|u\|_{L^\infty(\Omega)} + \|f\|_{L^p(\Omega)} \right).$$

The constant $C > 0$ depends only on d, λ, Λ, $\sup_{\alpha \in \mathcal{A}} \sup_{\beta \in \mathcal{B}} \|b_{\alpha,\beta}\|_{L^\infty(\Omega)}$, $\text{diam}(\Omega)$, and $\text{dist}(\Omega', \partial\Omega)$.

As discussed before, the general idea underlying the proof of Theorem 4.6 is to connect (4.5) with a Bellman operator of the form

$$\inf_{\beta \in \mathcal{B}} \left(- \text{Tr}(A_\beta M) \right), \tag{4.6}$$

since the homogeneous equations governed by (4.6) is concave. It follows from Corollary 4.1 that solutions to

$$\inf_{\beta \in \mathcal{B}} \left(- \text{Tr}\left(A_\beta D^2 h \right) \right) = 0 \quad \text{in} \quad \Omega$$

are locally of class $C^{2,\alpha}$, with estimates.

When establishing a regularity result in $W^{2,p}$-spaces, the usual assumption on the approximated model is $C^{1,1}$-estimates; see Caffarelli (1989) and Caffarelli and Cabré (1995). Here we choose to prove the result by accessing a $W^{2,q}$-estimate for $d < p < q$. The motivation for this choice is to detail the computations in this context, even though under Assumption 4.3 our limiting equation has $C^{2,\alpha}$-estimates.

In line with the previous observation, we emphasize that Assumption 4.3 can be weakened to include variable coefficients $A_\beta = A_\beta(x)$. In this case, a further condition on the Bellman equation governed by the family $(A_\beta(\cdot))_{\beta \in \mathcal{B}}$ is required. We formulate such a condition next.

Condition 4.7 (Estimates in $W^{2,q}$-spaces for the approximate problem)   Let $h \in C(\Omega)$ be an L^p-viscosity solution to

$$\inf_{\beta \in \mathcal{B}} \left(- \operatorname{Tr} \left(A_\beta(x) D^2 h \right) \right) = 0 \quad \text{in} \quad \Omega. \tag{4.7}$$

Then $h \in W^{2,q}_{\mathrm{loc}}(\Omega)$ and, for every $\Omega' \Subset \Omega$ there exists a universal constant $C > 0$ such that

$$\|h\|_{W^{2,q}(\Omega')} \le C \|h\|_{L^\infty(\Omega)},$$

where $C = C(d, \lambda, \Lambda, \operatorname{diam}(\Omega), \operatorname{dist}(\Omega', \partial\Omega))$.

The regularity required in Condition 4.7 is available under mild conditions on the family $(A_\beta(\cdot))_{\beta \in \mathcal{B}}$. Define

$$\vartheta(x, x_0) := \sup_{\beta \in \mathcal{B}} |A_\beta(x) - A_\beta(x_0)|;$$

by asking $\|\vartheta(\cdot, x_0)\|_{L^\infty(\Omega)} \ll 1$ for every $x_0 \in \Omega$, we can frame the problem in the context of Caffarelli and Cabré (1995, Theorem 7.1). To make the precise correspondence, one should set

$$F(x, M) := \inf_{\beta \in \mathcal{B}} \left(- \operatorname{Tr} \left(A_\beta(x) M \right) \right);$$

hence

$$\frac{|F(x, M) - F(x_0, M)|}{1 + \|M\|}$$

$$= \frac{\left| \inf_{\beta \in \mathcal{B}} \left(- \operatorname{Tr} \left(A_\beta(x) M \right) \right) - \inf_{\beta \in \mathcal{B}} \left(- \operatorname{Tr} \left(A_\beta(x_0) M \right) \right) \right|}{1 + \|M\|}$$

$$\le \frac{|A_\beta(x) - A_\beta(x_0)| \, \|M\|}{1 + \|M\|}.$$

As a consequence

$$\left(\fint_{B_r(x)} \sup_{M \in S(d)} \left(\frac{|F(x, M) - F(x_0, M)|}{1 + \|M\|} \right)^d dx \right)^{\frac{1}{d}} \le \left(\fint_{B_r(x)} \vartheta(x, x_0)^d dx \right)^{\frac{1}{d}}$$

$$\ll 1.$$

Hence, the smallness regimes in Caffarelli and Cabré (1995, Theorem 7.1, (7.5)) is satisfied, and the regularity required in Condition 4.7 is available for the Bellman equation with variable coefficients.

We continue with the proof of Theorem 4.6, which comprises three major steps. First we consider the Isaacs equation with no dependence on lower-order terms – i.e., the case $b_{\alpha,\beta} \equiv 0$ – and prove a $W^{2,p}$-regularity result. Then we examine the full equation and establish a $C^{1,\alpha}$-estimate. Finally, a reduction argument completes the proof.

Remark 4.8 As discussed before, the issue of local regularity in Ω can be framed in terms of the open balls of radius r. Fix $\Omega' \Subset \Omega$. For every $0 < r < 1$, it is possible to cover Ω' with a finite family of open balls $(B_r(x_i))_{i=1,\dots,N}$ such that $x_i \in \Omega'$ and $\overline{B_r(x_i)} \subset \Omega$, for every $i = 1, \dots, N$. For that reason, we continue by setting $\Omega := B_1$.

4.3.1 A Purely Second-Order Isaacs Equation

In light of Remark 4.8, we consider a purely second-order Isaacs equation in the unit ball:

$$\sup_{\alpha \in \mathcal{A}} \inf_{\beta \in \mathcal{B}} \left(- \operatorname{Tr} \left(A_{\alpha,\beta}(x) D^2 u \right) \right) = f \quad \text{in} \quad B_1. \qquad (4.8)$$

In this section we prove a $W^{2,p}$-estimate for the L^p-viscosity solutions to (4.8). This is the content of the next proposition.

Proposition 4.9 *Let $u \in C(B_1)$ be an L^p-viscosity solution to (4.8) for $p > d$, fixed. Suppose Assumptions 4.2, 4.3, and 4.5 are in force. Then, $u \in W^{2,p}_{loc}(B_1)$ and there exists $C > 0$ such that*

$$\|u\|_{W^{2,p}(B_{1/2})} \le C \left(\|u\|_{L^\infty(B_1)} + \|f\|_{L^p(B_1)} \right),$$

where $C = C(d, \lambda, \Lambda)$.

As before, we start with Lin's integral estimates. More precisely, we start with its inhomogeneous variant. For the sake of completeness, we state it next in terms of the sets A_t introduced in Definition 3.11.

Lemma 4.10 ($W^{2,\delta}$-estimate) *Let $u \in C(B_1)$ be an L^p-viscosity solution to (4.8). Suppose Assumptions 4.2 and 4.5 are in force. Then there exist universal constants $C > 0$ and $\mu > 0$ such that*

$$|A_t(u, B_1) \cap Q| \le C t^{-\mu}, \qquad (4.9)$$

for every $t > 0$, where Q denotes the d-dimensional cube of unitary length.

We continue with an approximation lemma.

Proposition 4.11 *Let $u \in C(B_1)$ be an L^p-viscosity solution to (4.8), for $p > d$. Suppose 4.2, 4.3, and 4.5 are in force. Fix $q > p$. There exists $h \in W^{2,q}(B_{7/8}) \cap C(\overline{B_{8/9}})$ satisfying*

$$\|u - h\|_{L^\infty(B_{7/8})} + \|\varphi(x)\|_{L^p(B_{7/8})} \le C_0 \left(\varepsilon_1^\tau + \|f\|_{L^p(B_1)} \right),$$

where $C_0 = C_0(d, \lambda, \Lambda, C, q, p)$ and $\tau = \tau(d, \lambda, \Lambda, C, q, p)$ are nonnegative constants, $\varepsilon_1 > 0$ is the constant in Assumption 4.3, and

$$\varphi(x) := f - \sup_{\alpha \in \mathcal{A}} \inf_{\beta \in \mathcal{B}} \left(- \operatorname{Tr} \left(A_{\alpha, \beta}(x) D^2 h(x) \right) \right).$$

Moreover, there exists $C > 0$ such that

$$\|h\|_{W^{2,q}(B_{7/8})} \leq C \|h\|_{L^\infty(B_1)}.$$

Proof We split the proof into three steps. First we refine the estimate for the Hessian of h by introducing a parameter $\delta \in (0, 1/2)$. Then we resort to Assumption 4.3 and estimate the L^p-norm of the Isaacs operator evaluated at the solution of the Bellman equation. Finally, we combined several estimates with a universal choice for some parameters and complete the proof.

Step 1. Let $h \in C(\overline{B}_{8/9})$ be the unique L^p-viscosity solution to

$$\begin{cases} \inf_{\beta \in \mathcal{B}} \left(- \operatorname{Tr} \left(A_\beta D^2 h \right) \right) = 0 & \text{in} & B_{8/9} \\ h = u & \text{on} & \partial B_{8/9}. \end{cases}$$

Because h solves an homoegenous equation governed by a concave operator, it follows from Corollary 4.1 that

$$\|h\|_{W^{2,q}(B_{7/8})} \leq C \|h\|_{L^\infty(B_1)},$$

for some universal constant C.

On the other hand, the Krylov–Safonov theory implies the existence of $\gamma \in (0, 1)$ such that $u \in C^\gamma \operatorname{loc}(B_1)$ and $h \in C^{\gamma/2}_{\operatorname{loc}}(B_{8/9})$; moreover, the estimates

$$\|u\|_{C^\gamma(\overline{B_{8/9}})} \leq C \left(\|u\|_{L^\infty(B_1)} + \|f\|_{L^p(B_1)} \right) \tag{4.10}$$

and

$$\|h\|_{C^{\frac{\gamma}{2}}(\overline{B_{8/9}})} \leq C \left(\|u\|_{L^\infty(B_1)} + \|f\|_{L^p(B_1)} \right), \tag{4.11}$$

are available.

Now, we shrink a bit further the domain and estimate the Hessian of h in such smaller ball. Start by fixing $\delta \in (0, 1/2)$ and taking a point $x_0 \in B_{8/9-\delta}$. It is clear that $B_\delta(x_0) \subset B_{8/9}$. Take $x_1 \in \partial B_\delta(x_0)$ and apply the $W^{2,q}$-estimates available for h to its translation $h(x) - h(x_1)$, inside $B_\delta(X_0)$. It follows that

$$\|D^2 h\|_{L^q(B_{\delta/2}(x_0))} \leq C\delta^{\frac{d-2q}{q}} \|h - h(x_1)\|_{L^\infty(\partial B_\delta(x_0))}$$

$$\leq C\delta^{\frac{d-2q}{q} + \frac{\gamma}{2}} \left(\|u\|_{L^\infty(B_1)} + \|f\|_{L^p(B_1)} \right),$$

since (4.11) implies

$$|h(y) - h(x_1)| \leq C \left(\|u\|_{L^\infty(B_1)} + \|f\|_{L^p(B_1)} \right) |y - x_1|,$$

for every $y \in \partial B_\delta(x_0)$. As a consequence, we recover

$$\left\| D^2 h \right\|_{L^q(B_{8/9-\delta})} \leq C \delta^{\frac{d-2q}{q} + \frac{\gamma}{2} - d} \left(\|u\|_{L^\infty(B_1)} + \|f\|_{L^p(B_1)} \right). \qquad (4.12)$$

Step 2. At this point we use Assumption 4.3; because h solves the homogeneous Bellman equation, for every $x_0 \in B_{8/9-\delta}$ we have

$$\left| \sup_{\alpha \in \mathcal{A}} \inf_{\beta \in \mathcal{B}} \left(-\operatorname{Tr}\left(A_{\alpha,\beta}(x_0) D^2 h(x_0) \right) \right) \right|$$

$$= \left| \sup_{\alpha \in \mathcal{A}} \inf_{\beta \in \mathcal{B}} \left(-\operatorname{Tr}\left(A_{\alpha,\beta}(x_0) D^2 h(x_0) \right) \right) - \inf_{\beta \in \mathcal{B}} \left(-\operatorname{Tr}\left(A_\beta D^2 h(x_0) \right) \right) \right|$$

$$\leq C \varepsilon_1 \left\| D^2 h(x_0) \right\|,$$

where the last inequality follows from Assumption 4.3. By combining the former inequality with (4.12), we get

$$\left\| \sup_{\alpha \in \mathcal{A}} \inf_{\beta \in \mathcal{B}} \left(-\operatorname{Tr}\left(A_{\alpha,\beta}(x) D^2 h(x) \right) \right) \right\|_{L^p(B_{8/9-\delta})} \leq C_1, \qquad (4.13)$$

once C_1 is defined as

$$C_1 := C \delta^{\frac{d-2q}{q} + \frac{\gamma}{2} - d} \varepsilon_1 \left(\|u\|_{L^\infty(B_1)} + \|f\|_{L^p(B_1)} \right).$$

Finally, we combine (4.10) and (4.11) to obtain

$$\|u - h\|_{L^\infty(B_{8/9-\delta})} \leq C \delta^{\frac{\gamma}{2}} \left(\|u\|_{L^\infty(B_1)} + \|f\|_{L^p(B_1)} \right). \qquad (4.14)$$

Step 3. In this last step we gather (4.13) and (4.14) and apply the maximum principle to get

$$\|u - h\|_{L^\infty(B_{8/9-\delta})} \leq C \delta^{\frac{\gamma}{2}} \left(\|u\|_{L^\infty(B_1)} + \|f\|_{L^p(B_1)} \right) + C \|f\|_{L^p(B_{8/9-\delta})}$$

$$+ C \varepsilon_1 \delta^{\frac{d-2q}{q} + \frac{\gamma}{2} - d} \left(\|u\|_{L^\infty(B_1)} + \|f\|_{L^p(B_1)} \right)$$

$$\leq \left(C \delta^{\frac{\gamma}{2}} + C \varepsilon_1 \delta^{\frac{d-2q}{q} + \frac{\gamma}{2} - d} \right) \left(\|u\|_{L^\infty(B_1)} + \|f\|_{L^p(B_1)} \right)$$

$$+ C \|f\|_{L^p(B_1)}.$$

To complete the proof we choose

$$\delta := \varepsilon_1^{\frac{q}{dq + 2q - d}} \quad \text{and} \quad \tau := \frac{\gamma q}{2(dq + 2q - d)}$$

and notice $B_{7/8} \subset B_{8/9-\delta}$. $\qquad \square$

The approximation result in Proposition 4.11 does not rely on the usual compactness strategy. Moreover, the smallness parameter $0 < \varepsilon_1 \ll 1$ remains yet to be chosen. This will be done in the next lemma, where we start to refine the decay rate involved in the proof of Lemma 4.10.

Lemma 4.12 *Let $u \in C(B_1)$ be an L^p-viscosity solution to (4.8). Suppose that Assumptions 4.2, 4.3, and 4.5 hold true. Suppose further that*

$$-|x|^2 \le u(x) \le |x|^2 \quad in \quad B_1 \setminus B_{6/7}.$$

There exist $M^* > 1$ and $\sigma \in (0, 1)$ such that

$$|G_{M^*}(u, B_1) \cap Q| \ge 1 - \sigma.$$

Proof We split the proof into two steps. The first one concerns a measure-decay estimate for the function h from Proposition 4.11. Our second step resorts to an auxiliary function v, which relates the measure decay for h with the same quantity for u.

Step 1. We start with the function h from Proposition 4.11 and extend it outside $B_{7/8}$ so that

$$h = u \quad in \quad B_1 \setminus B_{8/9}$$

and

$$\|u - h\|_{L^\infty(B_1)} = \|u - h\|_{L^\infty(B_{7/8})}.$$

Because h solves a Dirichlet problem with boundary data given by u, the maximum principle ensures $\|h\|_{L^\infty(B_{7/8})} \le 1$. As a consequence,

$$-2 - |x|^2 \le h(x) \le |x|^2 + 2 \quad in \quad B_1 \setminus B_{7/8}. \tag{4.15}$$

Also, we learn from Corollary 4.1 that $h \in W^{2,q}_{\mathrm{loc}}(B_{7/8})$. The latter builds upon (4.15) to produce information in the entire unit ball; in fact we conclude the existence of $C > 0$ such that

$$|A_t(h, B_1) \cap Q| \le Ct^{-q}. \tag{4.16}$$

Step 2. We proceed by introducing an auxiliary function. Let $v \colon B_1 \to \mathbb{R}$ be defined as

$$v(x) := \frac{u(x) - h(x)}{2C_1 \varepsilon_1^\tau}.$$

We have

$$\sup_{\alpha \in \mathcal{A}} \inf_{\beta \in \mathcal{B}} \left(- \mathrm{Tr}\left(A_{\alpha, \beta}(x) D^2 v\right)\right) \le \frac{1}{2C_1 \varepsilon_1^\tau}\left(f + \mathcal{P}^-(D^2 h)\right) \le \frac{f}{2C_1 \varepsilon_1^\tau};$$

a reversed inequality follows from a similar argument. As a conclusion, v falls within the scope of Lemma 4.10. Hence,

$$|A_t(u - h, B_1) \cap Q| \leq C\varepsilon_1^{\gamma \mu} t^{-\mu}.$$

Write $M^* := 2t$ to obtain

$$|A_{M^*}(u, B_1) \cap Q| \leq |A_{M^*/2}(u - h, B_1) \cap Q| + |A_{M^*/2}(h, B_1) \cap Q|$$

$$\leq C\varepsilon_1^{\tau \mu} \left(\frac{M^*}{2}\right)^{-\mu} + \overline{C}\left(\frac{M^*}{2}\right)^{-q}.$$

By choosing $0 < \varepsilon_1 \ll 1$ small enough and $M^* \gg 1$ sufficiently large we complete the argument. $\qquad\square$

Remark 4.13 (Universal choice of ε_1)   When choosing the parameter $0 < \varepsilon_1 \ll 1$ in the end of the previous proof, we (universally) determine the smallness regime required in Assumption 4.3.

Our next lemma is the same as Proposition 1.44 and Lemma 3.16. We include it here together with its proof for completeness.

Lemma 4.14   *Let $u \in C(B_1)$ be an L^p-viscosity solution to (4.8). Suppose that Assumptions 4.2, 4.3, and 4.5 hold true. Let Q^* be a cube such that $Q \subset Q^*$. If*

$$G_1(u, B_1) \cap Q^* \neq \emptyset,$$

then

$$|G_M(u, B_1) \cap Q| \geq 1 - \sigma,$$

for some $M > 1$.

Proof   Notice that, since $G_1(u, B_1) \cap Q^* \neq \emptyset$, one can find $x_0 \in B_1$ and an affine function $\ell(x)$ such that

$$-\frac{|x - x_0|^2}{2} \leq u(x) - \ell(x) \leq \frac{|x - x_0|^2}{2}$$

in B_1; it follows from the definition of G_1. Now, choose $C > 0$ to ensure

$$v(x) := \frac{(u - \ell)(x)}{C}$$

is such that $\|v\|_{L^\infty(B_1)} \leq 1$ and

$$-|x|^2 \leq v(x) \leq |x|^2 \qquad \text{in} \qquad B_1 \setminus B_{6/7}.$$

We conclude from Lemma 4.12 that

$$|G_{M^*}(v, B_1) \cap Q| \geq 1 - \sigma.$$

Set $M := CM^*$; then

$$|G_M(u, B_1) \cap Q| = |G_{CM^*}(u, B_1) \cap Q| = |G_{M^*}(v, B_1) \cap Q| \geq 1 - \sigma,$$

which finishes the proof.   □

In what follows, the constant M points to Lemma 4.14 and stands for $M :=$ CM^*. We continue by exploring a consequence of the Calderón–Zygmund cube decomposition.

Lemma 4.15 *Let $u \in C(B_1)$ be an L^p-viscosity solution to (4.8). Suppose Assumptions 4.2, 4.3, and 4.5 are in force. Extend f by zero outside of B_1 and define the sets*

$$A := A_{M^{k+1}}(u, B_1) \cap Q$$

and

$$B := \left(A_{M^k}(u, B_1) \cap Q\right) \cup \left\{x \in Q \mid m(f^p) \geq (CM^k)^p\right\}.$$

Then

$$|A| \leq \sigma |B|.$$

Proof The proof relies on the Calderón–Zygmund cube decomposition, as used before. We start by noticing an implication of Lemma 4.12. Indeed, it yields

$$\left|G_{M^{k+1}}(u, B_1) \cap Q\right| \geq |G_M(u, B_1) \cap Q| \geq \left|G_{C_3\overline{M}}(u, B_1) \cap Q\right| \geq 1 - \sigma.$$

In turn, the definition of A leads to

$$|A| \leq \sigma.$$

Set $K := Q_{1/2^i}(x_0)$ and denote with K^* the predecessor of K. If we verify that the equality

$$\left|A_{M^{k+1}}(u, B_1) \cap K\right| = |A \cap K| > \sigma |K| \tag{4.17}$$

implies the inclusion $K^* \subset B$, the proof is complete.

We argue by contradiction; suppose $K^* \not\subset B$. Then there exists x_1 such that

$$x_1 \in K^* \cap G_{M^k}(u, B_1) \tag{4.18}$$

and

$$m(f^p)(x_1) < \left(CM^k\right)^p. \tag{4.19}$$

Consider the affine map $T : Q \to K$, defined as

$$T(y) := x_0 + \frac{y}{2^i}.$$

We resort to the composition of u and T, normalized by a scaling factor. Define $u^* : Q \to \mathbb{R}$ as

$$u^*(y) := \frac{2^{2i}}{M^k}(u \circ T)(y) = \frac{2^{2i}}{M^k}u\left(x_0 + \frac{y}{2^i}\right).$$

Our goal is to verify that u^* falls within the scope of Lemma 4.14. Indeed,

$$D^2 u^*(y) = \frac{1}{M^k}D^2 u\left(x_0 + \frac{y}{2^i}\right);$$

hence

$$\sup_{\alpha \in \mathcal{A}} \inf_{\beta \in \mathcal{B}} \left(-\operatorname{Tr}\left(A_{\alpha,\beta}\left(x + \frac{y}{2^i}\right)D^2 u^*\right)\right) = f^* \quad \text{in} \quad B_1,$$

for a source term f^* given by

$$f^*(y) := \frac{f\left(x_0 + 2^{-i}y\right)}{M^k}.$$

A straightforward computation yields

$$\|f^*\|_{L^p(B_{2^{-i}})}^p = \frac{2^{i(2-d)}}{M^{kp}}\int_{B_{2^{-i}}(x_0)}|f(x)|^p dx \le \varepsilon_1^p,$$

where the last inequality follows from properties of the maximal function and (4.19). From (4.18) we obtain

$$G_1\left(u^*, B_{2^{-i}}(x_0)\right) \cap Q^* \ne \emptyset.$$

Hence,

$$\left|G_M\left(u^*, B_{2^{-i}}(x_0)\right) \cap Q\right| \ge (1 - \sigma)|Q|,$$

i.e.,

$$\left|G_{M^{k+1}}(u, B_1) \cap K\right| \ge (1 - \sigma)|K|;$$

the former inequality yields a contradiction with (4.17), which finishes the proof.   □

We are now in a position to complete the proof of Proposition 4.9.

Proof of Proposition 4.9 To complete the proof we need to ensure that there exists $C > 0$ for which

$$\sum_{k=1}^{\infty} M^{pk} \left| A_{M^k}(u, B_{1/2}) \right| \leq C. \qquad (4.20)$$

For $k \in \mathbb{N}$, define

$$a_k := \left| A_{M^k}(u, B_1) \cap Q \right|$$

and

$$b_k := \left| \{ x \in Q \mid m(f^p)(x) \geq (CM^k)^p \} \right|.$$

Because of Lemma 4.15, we have

$$a_{k+1} \leq a_k + b_k.$$

Hence,

$$a_k \leq \sigma^k + \sum_{i=0}^{k-1} \sigma^{k-i} b_i.$$

Since $f^q \in L^{p/q}(B_1)$ the maximal function satisfies $m(f^q) \in L^{p/q}(B_1)$ and

$$\left\| m(f^q) \right\|_{L^{p/q}(B_1)} \leq C \| f \|_{L^p(B_1)}^q \leq C,$$

for some $C > 0$. It allows us to bound

$$\sum_{k=0}^{\infty} M^{pk} b_k \leq C.$$

Therefore, we find

$$\sum_{k=1}^{\infty} M^{pk} a_k \leq \sum_{k=1}^{\infty} \left(\sigma M^p \right)^k + \sum_{k=1}^{\infty} \sum_{i=0}^{k-1} \sigma^{k-i} M^{p(k-i)} M^{pi} b_i$$

$$\leq \sum_{k=1}^{\infty} \left(\sigma C M^{*p} \right)^k + \sum_{k=1}^{\infty} \sum_{i=0}^{k-1} \sigma^{k-i} M^{p(k-i)} M^{pi} b_i$$

$$\leq \sum_{k=1}^{\infty} 2^{-k} + \left(\sum_{i=0}^{\infty} M^{pi} b_i \right) \left(\sum_{j=1}^{\infty} 2^{-j} \right)$$

$$\leq C,$$

and complete the proof. $\square$

We have proved a $W^{2,p}$-regularity result for a purely second-order Isaacs equation, approximating it with a Bellman operator. The next step towards the result to (4.5) is an estimate for the gradient of the solutions.

Working under Assumption 4.3 and the condition $p > d$, we resort to arguments in Święch (1997, Theorem 2.1) and prove that L^p-viscosity solutions to (4.5) are in $C_{\text{loc}}^{1,\alpha}(\Omega)$, for some $\alpha \in (0,1)$, with estimates. This fact enables us to reduce the analysis to the case treated in Proposition 4.9, taking into account a modified source term. Such modification is required to accommodate the gradient Du. Next, we detail the proof of $C^{1,\alpha}$-estimates for the solutions to (4.5).

4.3.2 Gradient Estimates for the Full Equation

In this section we recall an estimate for the gradient of the solutions to (4.5) in Hölder spaces; it appeared for the first time in Święch (1997). As before, we work in the unit ball B_1.

Theorem 4.16 (Estimates in $C^{1,\alpha}$-spaces) *Let $u \in C(B_1)$ be an L^p-viscosity solution to (4.5). Suppose Assumptions 4.2, 4.4, and 4.5 are in force, and $p > d$. Then there exists $\alpha \in (0,1)$ such that $u \in C_{\text{loc}}^{1,\alpha}(B_1)$. In addition,*

$$\|u\|_{C^{1,\alpha}(B_{1/2})} \leq C \left(\|u\|_{L^\infty(B_1)} + \|f\|_{L^p(B_1)} \right),$$

where $C > 0$ is a universal constant.

Remark 4.17 (Upper bound for the exponent $\alpha \in (0,1)$) The exponent $\alpha \in (0,1)$ in Theorem 4.16 satisfies a constraint depending on p and the dimension d. We have

$$\alpha \leq 1 - \frac{d}{p}.$$

It follows from scaling properties of the equation and Assumption 4.5; c.f. Theorem 1.56.

The proof of Theorem 4.16 follows closely the ideas detailed in Święch (1997, Theorem 2.1). We start with a variant of Proposition 4.11, obtained through a compactness argument. It accounts for an approximation lemma in the presence of lower order terms; see Święch (1997, Lemma 2.3).

Proposition 4.18 *Let $u \in C(B_1)$ be an L^p-viscosity solution to (4.5). Suppose Assumptions 4.2–4.5 are in force and $p > d$. For every $\delta > 0$, it is possible to choose $\varepsilon_1 > 0$ so that there exists $h \in W_{\text{loc}}^{2,q}(B_1)$ satisfying*

$$\|u - h\|_{L^\infty(B_{7/8})} \leq \delta.$$

In addition, there exists a universal constant $C > 0$ for which

$$\|h\|_{W^{2,q}(B_{1/2})} \le C \, \|h\|_{L^\infty(B_1)} \, .$$

Proof   As usual, we use a contradiction argument. By supposing the statement of the proposition is false, we find a sequence $(A_{\alpha,\beta}^n(\cdot))_{n\in\mathbb{N}}$ of (λ,Λ)-elliptic matrices, a sequence $(b_{\alpha,\beta}(\cdot)^n)_{n\in\mathbb{N}}$ of vector fields and sequences of functions $(u_n)_{n\in\mathbb{N}}$ and $(f_n)_{n\in\mathbb{N}}$ satisfying

$$\sup_{\alpha\in\mathcal{A}} \inf_{\beta\in\mathcal{B}} \left(- \operatorname{Tr} \left(A_{\alpha,\beta}^n(x) D^2 u_n \right) - b_{\alpha,\beta}^n(x) \cdot Du \right) = f_n \quad \text{in} \quad B_1, \quad (4.21)$$

with

$$\left| A_{\alpha,\beta}^n(x) - A_\beta \right| + \left\| b_{\alpha,\beta}^n \right\|_{L^\infty(B_1)} + \left\| f_n \right\|_{L^p(B_1)} < \frac{1}{n},$$

uniformly in $\alpha \in \mathcal{A}$ and $\beta \in \mathcal{B}$, such that

$$\|u_n - h\|_{L^\infty(B_{7/8})} > \delta_0$$

for some $\delta_0 > 0$ and every function $h \in W_{\text{loc}}^{2,q}(B_1)$ with

$$\|h\|_{W^{2,q}(B_{1/2})} \le C \, \|h\|_{L^\infty(B_1)} \, .$$

Local estimates in Hölder spaces are available for the solutions to (4.21); hence there exists $u_\infty \in C(B_1)$ such that $u_n \to u_\infty$ locally uniformly, through a subsequence if necessary. As a consequence, u_n converges, through a subsequence if necessary, to a function u_∞ in the C^ν-topology, for some $\nu \in (0,1)$; see Święch (1997, Lemma 1.9). Now, the stability of viscosity solutions leads to

$$\inf_{\beta\in\mathcal{B}} \left(- \operatorname{Tr} \left(A_\beta D^2 u_\infty \right) \right) = 0 \quad \text{in} \quad B_{8/9};$$

see Theorem 1.17; see also Caffarelli et al. (1996, Theorem 3.8) and Święch (1997, Lemma 1.7). The Evans–Krylov theory applies, implying u_∞ is of class $C^{2,\alpha}$. Of course, $u_\infty \in W_{\text{loc}}^{2,q}(B_1)$. Define $h := u_\infty$ to get a contradiction and complete the proof. ☐

The proof of Theorem 4.16 follows along the same lines as in Święch (1997, Theorem 2.1, Case 2). In fact the arguments leading to $W^{1,q}$-estimates in that paper also produce the conclusion of Theorem 4.16, provided the integrability of the source term is strictly above the dimension. We remark that, in Święch (1997), the approximation strategy is detailed in its full generality, relating a fully nonlinear operator $F = F(x,r,p,M)$ to its counterpart $F(0,0,0,M)$. In our case, we replace these functions with the Isaacs and Bellman operators, respectively. Alternatively, once Proposition 4.18 is available, Theorem 4.16 follows from the arguments leading to Theorem 1.56.

4.3.3 Sobolev Regularity for the Isaacs Equation

From Theorem 4.16, we learn that $|Du|$ is a bounded function. Hence

$$\tilde{f}(x) := \sup_{\alpha \in \mathcal{A}} \inf_{\beta \in \mathcal{B}} |b_{\alpha,\beta}(x)||Du| + f \in L^p(B_1).$$

The strategy to prove Theorem 4.6 is to reduce (4.5) to its simplified variant (4.8), not depending on the gradient, by resorting to $\tilde{f}$. We detail the argument in what follows.

Proof of Theorem 4.6 Because u is an L^p-viscosity solution to (4.5), we know that it is twice differentiable almost everywhere, and satisfies the equation for a.e.-$x \in B_1$; see Caffarelli et al. (1996, Theorem 3.6). As a consequence,

$$g(x) := \sup_{\alpha \in \mathcal{A}} \inf_{\beta \in \mathcal{B}} \left(-\operatorname{Tr}\left(A_{\alpha,\beta}(x)D^2u(x)\right)\right)$$

is defined almost everywhere in B_1. Now,

$$|g(x)| \leq \sup_{\alpha \in \mathcal{A}} \sup_{\beta \in \mathcal{B}} |b_{\alpha,\beta}(x)||Du(x)| + |f(x)| \in L^p(B_1),$$

because of Theorem 4.16. To complete the proof we notice that u solves

$$\sup_{\alpha \in \mathcal{A}} \inf_{\beta \in \mathcal{B}} \left[-\operatorname{Tr}\left(A_{\alpha,\beta}(x)D^2u(x)\right)\right] = g(x) \quad \text{in} \quad B_1$$

in the L^p-viscosity sense; see Crandall et al. (1996, Theorem 3.3) and Święch (1997, Corollary 1.6). Therefore, an L^p-viscosity solution to (4.5) happens to solve a purely second-order problem. The result readily follows from Proposition 4.9.   □

4.4 Regularity for the Isaacs Equation in Hölder Spaces

In this section we produce a regularity result for the Isaacs equation in $C^{2,\alpha}$-spaces. Once again we approximate the Isaacs operator with a Bellman equation. In this case, however, the proximity regime is much more stringent.

Next, we denote by $\gamma_0 \in (0,1)$ the exponent in the Evans–Krylov regularity theory associated with a (λ, Λ)-elliptic operator in dimension d. It follows that if, if $v \in C(\Omega)$ is an L^p-viscosity solution to

$$\inf_{\beta \in \mathcal{B}} \left(-\operatorname{Tr}\left(A_\beta D^2v\right)\right) = 0 \quad \text{in} \quad \Omega,$$

where $F: S(d) \to \mathbb{R}$ is a convex (λ, Λ)-elliptic operator, the Evans–Krylov regularity theory yields $v \in C^{2,\gamma_0}_{\text{loc}}(\Omega)$. We continue with our first assumption.

Assumption 4.19 (Smallness regime, estimates in $C^{2,\gamma}$) For $\gamma \in (0,\gamma_0)$ fixed arbitrarily, we suppose there exists $0 < \varepsilon_2 \ll 1$ for which

$$\sup_{x\in B_r} \sup_{\alpha\in\mathcal{A}} \sup_{\beta\in\mathcal{B}} \left|A_{\alpha,\beta}(x) - A_\beta\right| \le \varepsilon_2 r^\gamma,$$

for every $r > 0$ such that $B_r \subset \Omega$.

Moreover, we impose a condition on the integrability of the source term f. If we regard the L^p-norm of f over balls B_r as a function of $r > 0$, the next condition resembles a C^γ-type of estimate for that function.

Assumption 4.20 (Integrability of the source term) For $\gamma \in (0,\gamma_0)$ fixed arbitrarily, we suppose the source term $f \in L^p$ satisfies

$$\fint_{B_r} |f(x)|^p \mathrm{d}x \le \varepsilon_2^p r^{\gamma p}.$$

for every $r > 0$ such that $B_r \subset \Omega$.

Working under these assumption, we prove the following result.

Theorem 4.21 (Estimates in $C^{2,\gamma}$) *Let $u \in C(\Omega)$ be an L^p-viscosity solution to (4.5). Suppose Assumptions 4.2, 4.19, and 4.20 are in force and $p > d$. Let $\gamma \in (0,\gamma_0)$ be fixed. Then the function u is of class $C^{2,\gamma}$ at the origin.*

The main ingredient in the proof of Theorem 4.21 is a sequence of polynomials $(P_n)_{n\in\mathbb{N}}$, of the form

$$P_n(x) := a_n + \mathrm{b}_n \cdot x + \frac{1}{2}x^T C_n x,$$

with $P_0 \equiv P_{-1} \equiv 0$. We ensure this sequence satisfies

$$\inf_{\beta\in\mathcal{B}} \left(-\operatorname{Tr}\left(A_\beta C_n\right)\right) = 0,$$

$$\|u - P_n\|_{L^\infty(B_{\rho^n})} \le \rho^{n(2+\gamma)}$$

and

$$|a_n - a_{n-1}|_+ \rho^{n-1}|\mathrm{b}_n - \mathrm{b}_{n-1}| + \rho^{2(n-1)}\|C_n - C_{n-1}\| \le C\rho^{(n-1)(2+\gamma)},$$

for every $n \ge 0$ and some $0 < \rho \ll 1$; the latter is a universal constant chosen within the argument. This is the subject of the next proposition.

Proposition 4.22 *Let $u \in C(B_1)$ be an L^p-viscosity solution to (4.5). Suppose Assumptions 4.2, 4.19, and 4.20 are in force and $p > d$. Let $\gamma \in (0,\gamma_0)$ be fixed. There exists a sequence of polynomials $(P_n)_{n\in\mathbb{N}}$, of the form*

$$P_n(x) := a_n + b_n \cdot x + \frac{1}{2} x^T C_n x,$$

with $P_0 \equiv P_{-1} \equiv 0$, and a number $\rho \in (0, 1/2)$ satisfying

$$\inf_{\beta \in \mathcal{B}} \left(- \mathrm{Tr}\left(A_\beta C_n \right) \right) = 0, \tag{4.22}$$

$$\| u - P_n \|_{L^\infty(B_{\rho^n})} \le \rho^{n(2+\gamma)} \tag{4.23}$$

and

$$|a_n - a_{n-1}|_+ \rho^{n-1} |b_n - b_{n-1}| + \rho^{2(n-1)} \| C_n - C_{n-1} \| \le C \rho^{(n-1)(2+\gamma)}, \tag{4.24}$$

for every $n \ge 0$.

Proof For ease of presentation, we split the proof into four steps. As mentioned before, we resort to an induction argument. Notice the case $n = 0$ follows from $P_0 = P_{-1} = 0$ and the normalization assumption on u. We continue with the induction hypothesis, namely: Suppose the case $n = k$ has already been verified. We examine next the case $n = k + 1$.

Step 1. We introduce an auxiliary function $v_k \colon B_1 \to \mathbb{R}$ defined by

$$v_k(x) := \frac{(u - P_k)\left(\rho^k x\right)}{\rho^{k(2+\gamma)}}.$$

A straightforward computation ensures that v_k is an L^d-viscosity solution to

$$\frac{1}{\rho^{\gamma k}} \sup_{\alpha \in \mathcal{A}} \inf_{\beta \in \mathcal{B}} \left(- \mathrm{Tr}\left(A_{\alpha,\beta}\left(\rho^k x\right)\left(\rho^{\gamma k} D^2 v_k(x) + C_k\right)\right)\right) = f_k(x), \tag{4.25}$$

in B_{ρ^k}, with

$$f_k(x) := \frac{f\left(\rho^k x\right)}{\rho^{\gamma k}}.$$

Step 2. In the following, we observe that Proposition 4.18 is available for v_k. Denote $M_k := \rho^{\gamma k} M + C_k$ and compute

$$\left| \sup_{\alpha \in \mathcal{A}} \inf_{\beta \in \mathcal{B}} \left(- \mathrm{Tr}\left(A_{\alpha,\beta}\left(\rho^k x\right) M_k\right)\right) - \inf_{\beta \in \mathcal{B}} \left(- \mathrm{Tr}\left(A_\beta M_k\right)\right) \right|$$

$$\le \sup_{\alpha \in \mathcal{A}} \sup_{\beta \in \mathcal{B}} \left| \mathrm{Tr}\left(\left(A_{\alpha,\beta}\left(\rho^k x\right) - A_\beta\right) M_k\right) \right|$$

$$\le \sup_{x \in B_1} \sup_{\alpha \in \mathcal{A}} \sup_{\beta \in \mathcal{B}} \left| A_{\alpha,\beta}\left(\rho^k x\right) - A_\beta \right| \| \rho^{\gamma k} M + C_k \|$$

$$\le C \varepsilon_2 \rho^{\gamma k} \| \rho^{\gamma k} M + C_k \|$$

$$\le C \varepsilon_2 \rho^{\gamma k} \left(\| M \| + \| C_k \| \right),$$

where we have used Assumption 4.19 to obtain the third inequality. Because (4.24) has been checked for $n = k$, we have

$$\|C_k\| \leq \frac{C(1 - \rho^{k-1})}{1 - \rho} \leq \frac{C}{1 - \rho} \leq \tilde{C}.$$

As a consequence,

$$\frac{1}{\rho^{\gamma k}}\left|\sup_{\alpha \in \mathcal{A}} \inf_{\beta \in \mathcal{B}}\left(-\text{Tr}\left(A_{\alpha,\beta}(\rho^k x) M_k\right)\right) - \inf_{\beta \in \mathcal{B}}\left(-\text{Tr}\left(A_\beta M_k\right)\right)\right| \leq C\varepsilon_2(1 + \|M\|).$$

Also, Assumption 4.20 ensures

$$\|f_k\|^p_{L^p(B_1)} = \frac{1}{\rho^{\gamma k}}\int_{B_1}\left|f(\rho^k x)\right|^p dx = \frac{1}{\rho^{\gamma k}}\fint_{B_{\rho^{\gamma k}}}|f(y)|^p dy \leq \varepsilon_2^p. \quad (4.26)$$

Arguing as before, we fix $\delta > 0$ to be determined further. There exists a choice of $\varepsilon_2 = \varepsilon_2(\delta)$ ensuring the existence of $h \in C(B_{8/9})$ such that

$$\frac{1}{\rho^{\gamma k}}\inf_{\beta \in \mathcal{B}}\left(-\text{Tr}\left(A_\beta(\rho^{\gamma k}D^2 h(x) + C_k)\right)\right) = 0 \quad \text{in} \quad B_{8/9} \quad (4.27)$$

and agreeing with v_k on $\partial B_{8/9}$, with

$$\|v_k - h\|_{L^\infty(B_{8/9})} \leq \delta. \quad (4.28)$$

Further we determine δ, which depends only on universal quantities. Once this parameter is set, we also fix ε_2 in Assumption 4.19.

As concerns the regularity of h, we combine (4.22) and the induction hypothesis to conclude that

$$\inf_{\beta \in \mathcal{B}}\left(-\text{Tr}\left(A_\beta \frac{C_k}{\rho^{\gamma k}}\right)\right) = \frac{1}{\rho^{\gamma k}}\inf_{\beta \in \mathcal{B}}\left(-\text{Tr}\left(A_\beta C_k\right)\right) = 0. \quad (4.29)$$

In addition,

$$0 = \frac{1}{\rho^{\gamma k}}\inf_{\beta \in \mathcal{B}}\left(-\text{Tr}\left(A_\beta(\rho^{\gamma k}D^2 h + C_k)\right)\right)$$
$$= \inf_{\beta \in \mathcal{B}}\left(-\text{Tr}\left(A_\beta\left(D^2 h + \frac{C_k}{\rho^{\gamma k}}\right)\right)\right).$$

The Evans–Krylov regularity theory implies that L^p-viscosity solutions to

$$\inf_{\beta \in \mathcal{B}}\left(-\text{Tr}\left(A_\beta D^2 v\right)\right) = 0 \quad \text{in} \quad B_{8/9}$$

are in $C^{2,\gamma_0}_{\text{loc}}(B_1)$ and

$$\|v\|_{C^{2,\gamma_0}(B_{1/2})} \leq C\|v\|_{L^\infty(B_1)},$$

where $\gamma_0 \in (0,1)$ and $C > 0$ are universal constants. It then follows from (4.29) that $h \in C^{2,\gamma_0}_{\text{loc}}(B_1)$ with

$$\|h\|_{C^{2,\gamma_0}(B_{1/2})} \leq C \|h\|_{L^\infty(B_1)}.$$

Finally,

$$\left\| h - \left[h(0) + Dh(0) \cdot x + \frac{1}{2}x^T D^2 h(0)x \right] \right\|_{L^\infty(B_\rho)} \leq C\rho^{2+\gamma_0}. \qquad (4.30)$$

Step 3. At this point we gather (4.28) and (4.30) to obtain

$$\left\| v_k - \left[h(0) + Dh(0) \cdot x + \frac{1}{2}x^T D^2 h(0)x \right] \right\|_{L^\infty(B_\rho)} \leq \delta + C\rho^{2+\gamma_0}.$$

Set

$$\delta := \frac{\rho^{2+\gamma}}{2}, \qquad \text{and} \qquad \rho := \left(\frac{1}{2C} \right)^{\frac{1}{\gamma_0-\gamma}}$$

and

$$P_k^*(x) := h(0) + Dh(0) \cdot x + \frac{1}{2}x^T D^2 h(0)x;$$

then

$$\|v_k - P_k^*\|_{L^\infty(B_\rho)} \leq \rho^{2+\gamma}. \qquad (4.31)$$

Step 4. Next we define P_{k+1} and complete the induction argument. The definition of v_k builds upon (4.31) to produce

$$\sup_{x \in B_{\rho^{k+1}}} \left| u(x) - P_k(x) - \rho^{k(2+\gamma)} P_k^*(\rho^{-k}x) \right| \leq \rho^{(k+1)(2+\gamma)}.$$

By defining $P_{k+1}(x) := P_k(x) + \rho^{k(2+\gamma)} P_k^*(\rho^{-k}x)$, one obtains

$$\|u - P_{k+1}\|_{L^\infty(B_{\rho^{k+1}})} \leq \rho^{(k+1)(2+\gamma)};$$

it produces (4.23) for $n = k + 1$. Concerning (4.22), we observe that $C_{k+1} = C_k + \rho^{\gamma k} D^2 h(0)$; then (4.27) implies

$$\inf_{\beta \in \mathcal{B}} \left(-\text{Tr}\left(A_\beta C_{k+1} \right) \right) = 0.$$

Finally, to verify (4.24), we observe that

$$|a_{k+1} - a_k| \leq \rho^{k(2+\alpha)} h(0),$$

$$\rho^k |b_{k+1} - b_k| \le \rho^{k(2+\alpha)} Dh(0),$$

and

$$\rho^{2k} \|C_{k+1} - C_k\| \le \rho^{k(2+\alpha)} D^2 h(0).$$

To finish the proof it suffices to apply the uinversal C^{2,γ_0}-estimates available for h.   □

We are now in a position to prove Theorem 4.21.

Proof of Theorem 4.21 We argue as in Caffarelli and Cabré (1995, Chapter 8). Because of Proposition 4.22, the sequence $(P_n)_{n\in\mathbb{N}}$ converges to a second-order polynomial P^*, locally uniformly in B_1. The uniform estimates available for the coefficients of each P_n ensure that

$$|DP^*(0)| + \|D^2 P^*(0)\| \le C,$$

for some universal constant C. The uniform convergence also ensures that

$$\|u - P^*\|_{L^\infty(B_{\rho^n})} \le C\rho^{n(2+\gamma)},$$

which concludes the argument.   □

Bibliographical Notes

We have highlighted the importance of viscosity solutions in the context of the Isaacs equation. For the theory of viscosity solutions in the context of first-order Hamilton–Jacobi equations we refer the reader to the seminal works of Crandall and Lions (1983) and Crandall et al. (1984); see also Lions (1982). We also mention the monographs by Bardi and Capuzzo-Dolcetta (1997) and Barles (1994). For the connection of stochastic optimal control theory and viscosity solutions we mention the monograph by Fleming and Soner (2006).

As regards the connection of the Isaacs equations and the theory of viscosity solutions, we mention the works of Evans and Souganidis (1984), Fleming and Souganidis (1988, 1989), Lions and Souganidis (1988) and Jensen et al. (1988). See also the work of Święch (1996).

When it comes to the regularity available for the Isaacs equation we refer the reader to the results due to Kovats (2009a,b, 2012, 2016). For an Isaacs operator consisting of the minimum between a convex and a concave operator we mention the article by Cabré and Caffarelli (2003).

5

Regularity Theory for Degenerate Models

In this chapter we examine the regularity theory for certain classes of degenerate problems. The analysis is two-fold, as we cover the variational setting (*p*-Laplace model) and the nonvariational case (fully nonlinear operators). We start with a discussion on the *p*-Poisson equation.

5.1 A Regularity Theory for the *p*-Laplace Operator

The *p*-Laplace operator describes anisotropic processes with diffusivity depending on the gradient of the solutions. For functions u of class C^2 and $1 < p < \infty$ it is defined as

$$\Delta_p u := \operatorname{div}\left(|Du|^{p-2} Du\right).$$

The dependence of the coefficients on the solutions affects the ellipticity of the equation. At the points where the gradient vanishes, two possible regimes arise. Namely, for $1 < p < 2$ the equation becomes singular, whereas for $2 < p < \infty$ it degenerates.

Of particular interest are the extreme cases $p = 1$ and $p = \infty$. The first one yields

$$\Delta_1 u = \operatorname{div}\left(\frac{Du}{|Du|}\right)$$

and relates closely to geometric PDEs and the mean curvature tensor (and flow); see Evans and Spruck (1991, 1992a,b, 1995). The latter case yields the so-called infinity Laplacian

$$\Delta_\infty u := \langle D^2 u \cdot Du, Du\rangle.$$

151

This operator connects with the problem of optimal Lipschitz extensions. Although of fundamental importance to the general theory of PDEs, these cases are beyond the scope of our notes. We continue with a suitable definition of solution to the p-Poisson equation

$$\text{div}\left(|Du|^{p-2}Du\right) = f \qquad \text{in} \qquad B_1, \qquad (5.1)$$

where $f \in L^q_{\text{loc}}(B_1)$, for $q > 1$.

Definition 5.1 (Weak solution)   Let $1 < p < \infty$ and $u \in W^{1,p}_{\text{loc}}(B_1)$. We say that u is a weak (distributional) solution to (5.1) if

$$-\int_{B_1} |Du|^{p-2}(Du \cdot D\varphi)\mathrm{d}x = \int_{B_1} f\varphi\,\mathrm{d}x$$

for every $\varphi \in C^1_0(B_1)$.

The regularity theory for the solutions to (5.1) started in the context of homogeneous problems, $f \equiv 0$. In this setting, solutions are locally of class $C^{1,\alpha}$, for some $0 < \alpha < 1$, depending on p and the dimension, though unknown. These results were reported, independently, in Ural'ceva (1968) and Uhlenbeck (1977). A third proof appeared a few years later in Evans (1982b). As suggested by a class of examples, the Hölder continuity of the gradient is the optimal regularity for the solutions to (5.1).

The case of inhomogeneous equations has also been considered. In DiBenedetto (1983) the author proves that solutions to (5.1) are in $C^{1,\alpha}$, locally, where $0 < \alpha < 1$. Here, the exponent $\alpha = \alpha(d,p,q)$; i.e., the integrability of the source term impacts the regularity of the solutions. To see this, let $v(x) := |x|^\beta$. By computing its p-Laplacian we get

$$\Delta_p v = C_{\beta,p,d}\,|x|^{(\beta-1)(p-1)+\beta-2}.$$

By setting

$$\beta := 1 + \frac{1}{p-1},$$

we discover that the regularity of the solutions to (5.1) cannot be above $C^{1,\frac{1}{p-1}}$ if $f \in L^\infty$. On the other hand, by taking

$$\beta := \frac{d-pq}{(p-1)q},$$

we ensure $\Delta_p v \in L^q(B_1)$. As before, for $f \in L^q(B_1)$ it is not reasonable to expect the regularity of the solutions to (5.1) to be above $C^{1,\frac{d-pq}{(p-1)q}}$, locally.

The former observations give rise to (at least) two distinct questions on the regularity of the *p*-Poisson equation. The first one concerns the optimal regularity, provided $f \in L^{\infty}(B_1)$.

Conjecture 5.2 ($C^{p'}$-regularity conjecture) *Let* $u \in W_{\text{loc}}^{1,p}(B_1)$ *be a solution to* (5.1). *Suppose* $f \in L_{\text{loc}}^{\infty}(B_1)$. *Then* $u \in C_{\text{loc}}^{1,\frac{1}{p-1}}(B_1)$ *and there exists a constant* $C > 0$ *such that*

$$\|u\|_{C^{1,\frac{1}{p-1}}(B_{1/2})} \leq C \left(\|u\|_{L^{\infty}(B_1)} + \|f\|_{L^{\infty}(B_1)} \right).$$

The nomenclature $C^{p'}$-*regularity conjecture* is motivated by the form of the Hölder exponent

$$\frac{p-1}{p} + \frac{1}{p} = 1.$$

The proof of this conjecture in the planar case appeared in Araújo et al. (2017). A comprehensive discussion on the case of arbitrary dimension is the subject of Araújo et al. (2018). In parallel, the optimal regularity for the case $f \in L^q(B_1)$ can be conjectured as follows.

Conjecture 5.3 *Let* $u \in W_{\text{loc}}^{1,p}(B_1)$ *be a solution to* (5.1). *Suppose* $f \in L_{\text{loc}}^q(B_1)$. *Then* $u \in C_{\text{loc}}^{1,\frac{d-pq}{(p-1)q}}(B_1)$ *and there exists a constant* $C > 0$ *such that*

$$\|u\|_{C^{1,\frac{d-pq}{(p-1)q}}(B_{1/2})} \leq C \left(\|u\|_{L^{\infty}(B_1)} + \|f\|_{L^{\infty}(B_1)} \right).$$

In the case $d = 2$, this conjecture was established by Lindgren and Lindqvist (2017). Both conjectures are largely open in the case $d > 2$.

This section puts forward a strategy to study the regularity theory of the *p*-Poisson equation through approximation methods. We regard Δ_p as a perturbation of the Laplace operator. Since $\Delta_2 = \Delta$, if we consider any perturbation of the exponent $p = 2$, the equation enters the nonlinear realm and jeopardizes uniform ellipticity.

From a heuristic viewpoint, if $|p - 2| \ll 1$ and the source term f is bounded, the regularity regime of the *p*-Poisson equation would behave *as a perturbation* of the regularity theory available for $\Delta u = f$.

In what follows, we develop these arguments for $p \sim 2 + \varepsilon$. In this context, the approximation regime relates Δ_p to Δ. The result associated with this approach can be phrased along the following lines: Given $\alpha \in (0, 1)$, there exists $p_\alpha > 2$ such that if $p \in (2, p_\alpha)$, then solutions to (5.1) are locally $C^{1,\alpha}$-regular.

5.1.1 Transmiting Regularity from the Laplace Operator

We perturb the Laplace equation in a p-Poisson manner, passing from a homogeneous isotropic diffusion to a degenerate-elliptic, anisotropic, process. The analysis relies on an approximation strategy, relating the Δ_p operator with the Laplacian, by requiring the exponent p to be in a vicinity of 2. By means of such an approximation regime, the goal is to import information from the Laplace equation to the solutions of the p-Poisson problem.

The result we prove in this direction reads as follows.

Theorem 5.4 (Improved regularity for the p-Poisson equation) *Let $u \in W^{1,p}_{\mathrm{loc}}(B_1)$ be a bounded weak solution to*

$$\Delta_p u = f \qquad in \qquad B_1,$$

where $p > 2$ and $f \in L^\infty_{\mathrm{loc}}(B_1)$. For every $\alpha \in (0,1)$ there exists $p_\alpha > 2$, depending on α and the dimension, such that, if

$$2 < p < p_\alpha,$$

then $u \in C^{1,\alpha}_{\mathrm{loc}}(B_1)$. In addition, there exists a universal constant $C > 0$ for which

$$\|u\|_{C^{1,\alpha}(B_{1/2})} \leq C \left(\|u\|_{L^\infty(B_1)} + \|f\|_{L^\infty(B_1)} \right).$$

The former result appeared in Pimentel et al. (2020) and relies on several of ingredients: stability of solutions with respect to the exponent, the existence of harmonic correctors and a few oscillation controls.

We proceed by establishing a notion of stability for the weak solutions of $\Delta_p u = f$ with respect to the parameter p.

Proposition 5.5 (Sequential stability of weak solutions) *Let $(u_n)_{n\in\mathbb{N}} \subset W^{1,p}(B_1)$ and $(f_n)_{n\in\mathbb{N}} \subset L^q(B_1)$ be sequences of functions, with $q > 1$. Let $(p_n)_{n\in\mathbb{N}} \subset \mathbb{R}$ be a sequence of real numbers. Suppose*

$$\mathrm{div}\left(|Du_n|^{p_n-2} Du_n \right) = f_n \qquad in \qquad B_1, \tag{5.2}$$

and

$$|p_n - 2| + \|f_n\|_{L^q(B_1)} \leq \frac{1}{n}.$$

Let $u_\infty \in C^1(B_1)$ be such that u_n converges to u_∞ in $C^1(B_1)$. Then u_∞ is a weak solution to

$$\Delta u_\infty = 0 \qquad in \qquad B_{9/10}.$$

Proof To establish the proposition, it suffices to verify that

$$\int_{B_1} Du_\infty \cdot D\varphi \, dx = 0$$

for every $\varphi \in C_0^1(B_1)$. We compute

$$\left| \int_{B_{9/10}} Du_\infty \cdot D\varphi \, dx \right| \leq \int_{B_1} \left(|D\varphi| \left| Du_\infty - |Du_n|^{p_n-2} Du_n \right| \right) dx$$

$$+ \int_{B_1} |f_n||\varphi| dx$$

$$\leq C \int_{B_1} \left| Du_\infty - |Du_n|^{p_n-2} Du_n \right| dx + C \, \|f_n\|_{L^q(B_1)}.$$

The regularity theory available for the p-Poisson equation ensures

$$u_n \in C_{\text{loc}}^{1,\beta}(B_1) \quad \text{and} \quad \|u_n\|_{C^{1,\beta}(B_{9/10})} \leq C, \tag{5.3}$$

for some $\beta \in (0,1)$ and some $C > 0$; see DiBenedetto (1983). We observe the information in (5.3) is uniform in $n \in \mathbb{N}$. That is, β and C do not depend on $n \in \mathbb{N}$; see, for instance, Ladyzhenskaya and Ural'tseva (1968, Chapter 2). Moreover,

$$\left| Du_\infty - |Du_n|^{p_n-2} Du_n \right| \leq \|Du_\infty\|_{L^\infty(B_1)} + |Du_n|^{p_n-1} \leq C.$$

In addition, $|Du_n|^{p_n-2} Du_n \to Du_\infty$ a.e.-$x \in B_1$ as $n \to \infty$; hence, the Lebesgue Dominated Convergence Theorem yields

$$\left| \int_{B_1} Du_\infty \cdot D\varphi \, dx \right| = 0,$$

which finishes the proof. $\square$

As before, we rely on a localization argument, operating through scaling. Since the $C^{1,\alpha}$-scaling involves comparison with an affine function $\ell(x)$, we are interested in the equation solved by

$$v(x) := \frac{u(\rho x) - \ell(\rho x)}{\rho^{1+\alpha}}.$$

However, we notice that v solves a different equation, which places the analysis off the scope of the approximation methods discussed so far. The strategy to circumvent this roadblock is the introduction of small correctors; see Araújo et al. (2017). In the present setting, such structures arise from the connection of the solutions to $\Delta_p u = f$ with harmonic functions, provided $p - 2 \ll 1$.

Proposition 5.6 (Existence of C^1-small correctors) *Let $u \in W_{loc}^{1,p}(B_1)$ be a bounded weak solution to*

$$\Delta_p u = f \qquad in \qquad B_1,$$

where $p > 2$ and $f \in L^q(B_1)$, for $q > 1$. Given $\delta > 0$, there exists $\varepsilon > 0$ such that, if

$$(p - 2) + \|f\|_{L^q(B_1)} \leq \varepsilon,$$

then one can find $\xi \in C^1(B_{9/10})$ satisfying

$$|\xi(x)| < \delta \qquad and \qquad |D\xi(x)| < \delta$$

for $x \in B_{9/10}$, with

$$\Delta(u + \xi) = 0 \qquad in \qquad B_{9/10}.$$

Proof In line with the reasoning in the former approximation lemmas, we prove the proposition through a contradiction argument. We split the proof into three steps.

Step 1. Suppose its statement is false. Then there are sequences of functions $(u_n)_{n\in\mathbb{N}}$ and $(f_n)_{n\in\mathbb{N}}$ and a sequence of real numbers $(p_n)_{n\in\mathbb{N}}$ such that

$$\text{div}\left(|Du_n|^{p_n-2}Du_n\right) = f_n \qquad in \qquad B_1,$$

and

$$(p_n - 2) + \|f_n\|_{L^q(B_1)} < \frac{1}{n},$$

but, for every $\xi \in C^1(B_{9/10})$ with

$$\Delta(u_n + \xi) = 0 \qquad in \qquad B_{9/10},$$

we have either

$$|\xi(x)| \geq \delta_0 \qquad or \qquad |D\xi(x)| \geq \delta_0,$$

for some $x \in B_{9/10}$ and some $\delta_0 > 0$.

Step 2. The regularity for the p-Poisson equation ensures that $(u_n)_{n\in\mathbb{N}}$ is equibounded in $C^{1,\beta}(B_{9/10})$, for some $\beta \in (0,1)$; notice both β and the bounds for the sequence are universal. Hence, one can find $u_\infty \in C^{1,\gamma}(B_{9/10})$, for every $0 < \gamma < \beta$, so that $u_n \to u_\infty$ in $C^{1,\gamma}(B_{9/10})$, up to a subsequence, if necessary.

Step 3. The sequential stability of weak solutions in Proposition 5.5 ensures that

$$\Delta u_\infty = 0 \quad \text{in} \quad B_{9/10}.$$

Define $\xi_n := u_\infty - u_n$. We conclude

$$\Delta (u_n + \xi_n) = 0 \quad \text{in} \quad B_{9/10}$$

and

$$|\xi_n(x)| + |D\xi_n(x)| < \delta_0,$$

for every $x \in B_{9/10}$ and $n \gg 1$. Hence, we have reached a contradiction and the proof is complete. □

An interesting exercise is to compare the existence of small correctors with the previous approximation lemmas. The role of ξ is to *modify* the solutions to a p-Poisson equation and yield a harmonic function. A heuristic interpretation of Proposition 5.6 is the following: provided p is close to 2, solutions to the p-Poisson equation differ from a harmonic function by a well-defined, C^1-regular quantity.

Once C^1-small correctors are available, we are in a position to start examining the oscillation of the solutions to $\Delta_p u = f$. We proceed with the following proposition.

Proposition 5.7 *Let $u \in W^{1,p}(B_1)$ be a bounded weak solution to*

$$\Delta_p u = f \quad \text{in} \quad B_1,$$

where $p > 2$ and $f \in L^q(B_1)$. Take $\alpha \in (0,1)$, arbitrarily. There exist $\varepsilon_0 > 0$ and $0 < \rho \ll 1$ such that, if

$$|p - 2| + \|f\|_{L^q(B_1)} < \varepsilon_0, \tag{5.4}$$

then

$$\sup_{B_\rho} |u(x) - u(0)| \le \rho^{1+\alpha} + |Du(0)| \, \rho.$$

Proof We start by fixing $\delta > 0$, which is set further in the proof. Suppose $\varepsilon_0 > 0$ in (5.4) is chosen so that Proposition 5.6 yields a small corrector $\xi \in C^1(B_1)$ satisfying

$$|\xi(x)| < \frac{\delta}{3} \quad \text{and} \quad |D\xi(x)| < \frac{\delta}{3}$$

in $B_{9/10}$, with $u + \xi$ harmonic in $B_{9/10}$. For $x \in B_\rho$, the triangular inequality produces

$$|u(x)-[u(0)+Du(0)\cdot x]| \le |(u+\xi)(x)-[(u+\xi)(0)+D(u+\xi)(0)\cdot x]|$$
$$+|\xi(x)|+|\xi(0)|+|D\xi(0)\cdot x|$$
$$\le C\rho^2 + \delta. \tag{5.5}$$

We stress that $C > 0$ in (5.5) depends only on the dimension. In fact, because $u+\xi$ is harmonic in $B_{9/10}$, we recover

$$\|u+\xi\|_{C^2(B_{9/10})} \le C\big(\|u\|_{L^\infty(B_1)}+\|\xi\|_{L^\infty(B_{9/10})}\big) \le C,$$

where the second inequality follows from $\delta < 1$ and the fact that u is a bounded solution. By choosing

$$\rho := \left(\frac{1}{2C}\right)^{\frac{1}{1-\alpha}} \qquad \text{and} \qquad \delta := \frac{\rho^{1+\alpha}}{2}$$

we complete the argument. $\qquad\qquad\qquad\qquad\qquad\qquad\qquad\qquad\qquad\qquad\square$

Notice the choices of ρ and δ in the proof of Proposition 5.7 are universal; these choices relate the exponent $\alpha \in (0,1)$ to the ε-smallness regime required for the existence of small correctors in Proposition 5.6. The oscillation control obtained for a fixed $\rho > 0$ is consequential on the approximation regime. At this point the geometry of the PDE allows us to extend the oscillation control to arbitrary small scales.

For ease of notation, we present the next results in a vicinity of the origin. A simple translation $x \mapsto x+x_0$ recovers our conclusions to any point $x_0 \in B_{1/2}$, for any $0 < r < 1$ such that $B_r(x_0) \Subset B_1$.

Proposition 5.8 (Gradient-dependent control of oscillation) *Let* $u \in W^{1,p}(B_1)$ *be a bounded weak solution to*

$$\Delta_p u = f \qquad in \qquad B_1,$$

where $p > 2$ *and* $f \in L^q(B_1)$*. Take* $\alpha \in (0,1)$*, arbitrarily, satisfying*

$$q > \frac{d}{1-\alpha}. \tag{5.6}$$

There exist $\varepsilon_0 > 0$ *such that, if*

$$|p-2|+\|f\|_{L^q(B_1)} < \varepsilon_0, \tag{5.7}$$

then

$$\sup_{B_r} |u(x)-u(0)| \le Cr^{1+\alpha}\left(1+|Du(0)|r^{-\alpha}\right),$$

for some $C > 0$ *and every* $0 < r \ll 1$.

Proof We split the proof into three steps. First, we produce an oscillation control at discrete scales $n \in \mathbb{N}$. Then we extrapolate this estimate to the continuous setting.

Step 1. We establish

$$\sup_{B_{\rho^n}} |u(x) - u(0)| \leq \left(\rho^{n(1+\alpha)} + |Du(0)| \, \rho^n \right), \tag{5.8}$$

for every $n \in \mathbb{N}$, through an induction argument. The statement of Proposition 5.7 amounts to the case $n = 1$. Suppose the case $n = k$ has been verified. We prove the case $n = k + 1$.

Consider $v_k \colon B_1 \to \mathbb{R}$ defined as

$$v_k(x) := \frac{u(\rho^k x) - u(0)}{\rho^{k(1+\alpha)} + |Du(0)| \, \rho^k}.$$

The induction hypothesis implies that $\|v_k\|_{L^\infty(B_1)} \leq 1$. Furthermore,

$$Dv_k(x) = \frac{\rho^k Du(\rho^k x)}{\rho^{k(1+\alpha)} + |Du(0)| \, \rho^k} \quad \text{and} \quad Dv_k(0) = \frac{\rho^k Du(0)}{\rho^{k(1+\alpha)} + |Du(0)| \, \rho^k}.$$

Moreover, v_k is a weak solution to

$$\operatorname{div} \left(|Dv_k(x)|^{p-2} Dv_k(x) \right) = f_k(x) \quad \text{in} \quad B_1,$$

where

$$f_k(x) := \frac{\rho^{kp} f(\rho^k x)}{\left(\rho^{k(1+\alpha)} + |Du(0)| \, \rho^k \right)^{p-1}}. \tag{5.9}$$

Step 2. In the following we examine the integrability of f_k, as we must have $f_k \in L^q(B_1)$ for every $k \in \mathbb{N}$. To ensure this condition is met we need to impose an upper bound on the exponent p. Indeed, we have that $f_k \in L^q(B_1)$ provided

$$p - 2 \leq \frac{q - d}{\alpha q} - 1.$$

Since $p - 2 > 0$, we must have

$$q > \frac{d}{1 - \alpha}. \tag{5.10}$$

Set

$$\varepsilon^* := \min \left\{ \varepsilon_0, \frac{q - d}{\alpha q} - 1 \right\}$$

and impose

$$p - 2 \leq \frac{1}{2}\varepsilon^*;$$

then, we can apply Proposition 5.7 to v_k. It follows that

$$\sup_{B_\rho} |v_k(x) - v_k(0)| \leq \rho^{1+\alpha} + |Dv_k(0)|\, \rho. \qquad (5.11)$$

The definition of v_k and the estimate in (5.11) yields

$$\sup_{B_{\rho^{k+1}}} |u(x) - u(0)| \leq \rho^{(k+1)(1+\alpha)} + |Du(0)|\, \rho^{k+1+\alpha} + |Du(0)|\, \rho^{k+1},$$

$$(5.12)$$

by a simple scaling to B_1. Since $\rho^\alpha < 1$, we obtain (5.8).

Step 3. In what follows we consider $0 < r \ll 1$. Take $k \in \mathbb{N}$ such that $\rho^{k+1} \leq r \leq \rho^k$. Then

$$
\begin{aligned}
\sup_{B_r} \frac{|u(x) - u(0)|}{r^{1+\alpha}} &\leq \sup_{B_{\rho^k}} \frac{|u(x) - u(0)|}{\rho^{(k+1)(1+\alpha)}} \\
&\leq \frac{C\left(\rho^{k(1+\alpha)} + |Du(0)|\, \rho^k\right)}{\rho^{(k+1)(1+\alpha)}} \\
&\leq \frac{C}{\rho^{1+\alpha}}\left(1 + \frac{|Du(0)|}{\rho^{k\alpha}}\right) \\
&\leq C_\alpha \left(1 + |Du(0)|\, r^{-\alpha}\right),
\end{aligned}
$$

with

$$C_\alpha := \frac{C}{\rho^{1+\alpha}}.$$

The constant C_α depends only on universal quantities and the parameter $\alpha \in (0, 1)$, which is fixed. $\qquad\square$

We notice that Proposition 5.8 controls the oscillation of the function in the regions where the gradient is small. In fact, if $|Du(0)| < r^\alpha$, we get

$$\sup_{B_r} |u(x) - u(0)| \leq Cr^{1+\alpha}\left(1 + |Du(0)|\, r^{-\alpha}\right) \leq 2Cr^{1+\alpha}.$$

We continue with a proposition.

Proposition 5.9 (Oscillation at degenerate points)　*Let $u \in W^{1,p}(B_1)$ be a bounded weak solution to*

$$\Delta_p u = f \qquad in \qquad B_1,$$

where $p > 2$ and $f \in L^q(B_1)$. Take $\alpha \in (0,1)$, arbitrarily, satisfying (5.10). For $|Du(0)| < r^\alpha \ll 1$, there exists $C > 0$ such that

$$\sup_{B_r} |u(x) - u(0)| \le Cr^{1+\alpha}.$$

Proof The condition $|Du(0)| < r^\alpha$, combined with Proposition 5.8 leads to

$$\sup_{B_r} |u(x) - u(0)| \le Cr^{1+\alpha} \left(1 + |Du(0)| r^{-\alpha}\right) \le Cr^{1+\alpha},$$

and the proof is complete. □

On the other hand, the oscillation control in the regions where $|Du(0)| > r^\alpha$ should profit from the uniform ellipticity of the equation at the origin. Although heuristic in nature, this argument can be made rigorous.

Suppose that $|Du(0)| \ge r^\alpha$ and let $\lambda := |Du(0)|^{1/\alpha}$. Define $v \colon B_1 \to \mathbb{R}$ by

$$v(x) := \frac{u(\lambda x) - u(0)}{\lambda^{1+\alpha}}.$$

Notice first that $v(0) = 0$. Also,

$$|Dv(0)| = 1, \tag{5.13}$$

and

$$\operatorname{div}\left(|Dv(x)|^{p-2} Dv(x)\right) = \frac{\lambda^p f(\lambda x)}{\lambda^{(1+\alpha)(p-1)}} \quad \text{in} \quad B_1. \tag{5.14}$$

The regularity theory available for (5.14) implies Dv is continuous; hence (5.13) guarantees that $|Dv(x)| > 1/2$ in B_μ, for some $\mu > 0$, perhaps small. We conclude v solves a uniformly elliptic equation with bounded right-hand side in B_μ. It follows that $v \in C^{1,\alpha}(B_{9\mu/10})$, for every $\alpha \in (0,1)$. As a consequence, for every $\alpha \in (0,1)$ we have

$$\sup_{B_r} |v(x) - [v(0) + Dv(0) \cdot x]| \le Cr^{1+\alpha}, \tag{5.15}$$

for every $0 < r \ll 9\mu/10$ and some universal constant $C > 0$. The definition of v combined with (5.15) leads to

$$\sup_{B_r} \left| \frac{u(\lambda x) - u(0)}{\lambda^{1+\alpha}} - \frac{\lambda Du(0) \cdot x}{\lambda^{1+\alpha}} \right| \le Cr^{1+\alpha}.$$

That is,

$$\sup_{B_r} |u(x) - [u(0) + Du(0) \cdot x]| \le Cr^{1+\alpha},$$

for $0 < r < 9\lambda\mu/10$. Otherwise, if $9\lambda\mu/10 \le r < \lambda$, we get

$$\sup_{B_r} |u(x) - [u(0) + Du(0) \cdot x]| \le \sup_{B_\lambda} |u(x) - [u(0) + Du(0) \cdot x]|$$

$$\le \sup_{B_\lambda} |u(x) - u(0)| + |Du(0)|\,\lambda$$

$$\le (C + 1)\,\lambda^{1+\alpha}$$

$$\le (C + 1) \left(\frac{10r}{9\mu}\right)^{1+\alpha}$$

$$\le C r^{1+\alpha}.$$

The third inequality in the previous computation follows from a Taylor expansion for u together with elementary facts.

Proof of Theorem 5.4   We combine Proposition 5.9 and (5.16) to establish the result.   □

Remark 5.10 (Regularity conjectures)   The regularity conjectures discussed previously in this chapter have been established for $d = 2$. The case $f \in L^\infty(B_1)$ amounts to Conjecture 5.2, and its proof on the planar setting is the subject of Araújo et al. (2017). Conjecture 5.3 has also been established in the planar context; its proof is reported by Lindgren and Lindqvist (2017). The methods detailed above are similar to, and inspired by, ideas put forward in those papers. We believe that a solid understanding of the former computations may equip the reader to follow through the arguments in Araújo et al. (2017) and Lindgren and Lindqvist (2017) smoothly.

The result of Theorem 5.4 suggests that a degenerate perturbation of the Laplace operator resonates in the regularity of the solutions. However, the smaller the perturbation, the more regular are the solutions. In the following we frame this problem in the nonvariational, fully nonlinear setting. That is, we consider equations of the form

$$|Du|^\theta F\big(D^2u\big) = 1 \qquad \text{in} \qquad B_1, \tag{5.16}$$

where F is a (λ, Λ)-elliptic operator and the degeneracy exponent satisfies a constraint of the form $\theta > 0$. We discuss how the regularity of $F = 0$ transmits to the solutions to (5.16) in Section 5.2.

5.2 Fully Nonlinear Degenerate Problems

We now consider the regularity of the viscosity solutions to degenerate fully nonlinear equations of the form

$$|Du|^\theta F(D^2u) = f \qquad \text{in} \qquad B_1, \qquad (5.17)$$

where $f \in L^\infty(B_1)$ and $\theta > 0$.

Motivation for the study of elliptic problems modeled as in (5.17) comes from the so-called structure conditions. In fact, we would like to generalize Definition 1.12 to accommodate operators $G = G(p, M)$ satisfying growth conditions of the form

$$\sigma(|p|)\mathcal{P}^-(M - N) \le G(p, M) - G(p, N) \le \sigma(|p|)\mathcal{P}^+(M - N),$$

where $\sigma : \mathbb{R}^+ \to \mathbb{R}^+$ is a modulus of continuity. Clearly, by setting $\sigma(t) := t^\theta$ the problem in (5.17) satisfies the former structure condition.

In the regions where the norm of the gradient is uniformly bounded away from zero, (5.17) is uniformly elliptic. Clearly, where the gradient vanishes, ellipticity degenerates. The regularity theory for (5.17) starts with the Hölder continuity of the solutions. This fact unlocks compactness, which in turn builds upon stability of the solutions to produce an approximation result. At this level, it is of paramount relevance to notice that a solution to

$$|Du|^\theta F(D^2u) = 0 \qquad \text{in} \qquad B_1$$

solves

$$F(D^2u) = 0 \qquad \text{in} \qquad B_1 \qquad (5.18)$$

in the viscosity sense. Hence, (5.17) connects to a homogeneous uniformly elliptic equation. As a result, we are capable of transmitting a regularity theory in spaces $C^{1,\beta}$ from (5.18) to (5.17). A very interesting aspect is the role played by the degeneracy degree of the equation. In fact, the optimal regularity can be explicitly characterized in terms of θ.

Consider the example $v(x) := |x|^{1+\beta}$. Notice that

$$|Dv|^\theta \Delta v = C_\beta |x|^{\beta\theta+\beta-1};$$

if we set

$$\beta := \frac{1}{1+\theta}$$

it follows that $|Dv|^\theta \Delta v \in L^\infty(B_1)$. Therefore, the optimal regularity for the solutions to (5.17) cannot be above the class $C^{1,\beta}_{\text{loc}}(B_1)$. In the case of a general fully nonlinear operator, a distinctive feature of this class of problems is the trade-off between ellipticity and degeneracy.

Let $\alpha_0 \in (0, 1)$ denote the exponent associated with the C^{1,α_0}-regularity of the solutions to $F = 0$. The regularity theory for (5.17) arises from a competition between the degeneracy degree θ and the ellipticity, represented here by the exponent α_0. In fact, the findings reported in Imbert and Silvestre (2013) can be phrased as the following theorem.

Theorem 5.11 (Hölder-regularity of the gradient) *Let $u \in C(B_1)$ be a viscosity solution to*

$$|Du|^\theta F\left(D^2 u\right) = f \qquad in \qquad B_1,$$

where F is a (λ, Λ)-elliptic operator and $f \in L^\infty(B_1) \cap C(B_1)$. For every $\alpha \in (0,1)$ satisfying

$$\alpha \in (0, \alpha_0) \qquad and \qquad \alpha \leq \frac{1}{1+\theta},$$

we have $u \in C^{1,\alpha}_{\text{loc}}(B_1)$. In addition, there exists a universal constant $C > 0$ such that

$$\|u\|_{C^{1,\alpha}(B_{1/2})} \leq C \left(\|u\|_{L^\infty(B_1)} + \|f\|_{L^\infty(B_1)}^{\frac{1}{1+\theta}} \right).$$

In the case where F is convex/concave, the Evans–Krylov theory becomes available. In this setting, the sharp regularity of the solutions to (5.17) is completely driven by the degeneracy degree θ; in fact, in this case, $u \in C^{1,\frac{1}{1+\theta}}_{\text{loc}}(B_1)$; see Araújo et al. (2015). In the following, we detail the proof of this result.

5.2.1 A Cancellation Lemma

As previously mentioned, we rely on the compactness of the solutions for a perturbation of (5.17), an approximation result, and the fact that solutions to $|Du|^\theta F(D^2 u) = 0$ are solutions to $F(D^2 u) = 0$. We start by establishing the latter. If stated in a pedestrian fashion, the next proposition tells us that in the homogeneous setting, one can divide both sides of the equation by $|Du|^\theta$. This result first appeared in Imbert and Silvestre (2013, Lemma 6).

Proposition 5.12 (Cancellation lemma) *Let $u \in C(B_1)$ be a viscosity solution to the homogeneous equation*

$$|Du + q|^\theta F\left(D^2 u\right) = 0 \qquad in \qquad B_1,$$

where F is a (λ, Λ)-elliptic operator, $\theta > 0$ and $q \in \mathbb{R}^d$ is arbitrary. Then u also solves the uniformly elliptic problem

$$F\left(D^2 u\right) = 0 \qquad in \qquad B_1,$$

in the viscosity sense.

Proof For ease of presentation, we split the proof into four steps. We start by considering a reduction argument.

Step 1. Set $w(x) := u(x) + q \cdot x$. Clearly, w solves

$$|Dw|^{\theta} F(D^2 w) = 0 \qquad \text{in} \qquad B_1.$$

Since $D^2 w = D^2 u$, by proving the result for the previous equation we infer $F(D^2 u) = F(D^2 w)$ and complete the proof. Therefore, in what follows we suppose $q = 0$.

Step 2. Consider a test function $\varphi \in C^2(B_1)$ touching u from below at $x_0 \in B_1$. We suppose $u(x_0) = |x_0| = 0$ and

$$\varphi(x) := p \cdot x + \frac{1}{2} x^t M x,$$

for some $p \in \mathbb{R}^d$ and $M \in \mathbb{R}^{d^2}$. The equation for u yields

$$|p|^{\theta} F(M) \leq 0;$$

if $p \neq 0$ we obtain $F(M) \leq 0$ and the proof is complete. We then suppose $p \equiv 0$.

A further reduction concerns the spectrum of the matrix M. Suppose all the eigenvalues of M are nonpositive. In this case, ellipticity would yield $F(M) \leq 0$ and the proof would follow. Hence we suppose M has at least one (strictly) positive eigenvalue.

Step 3. We collect in $\{\sigma_1, \ldots, \sigma_k\}$ the eigenvectors associated with the k strictly positive eigenvalues of M. Set $E := \operatorname{span}\{\sigma_1, \ldots, \sigma_k\}$ and consider $\mathbb{R}^d =: E \oplus G$. Denote with P_E the orthogonal projection on E and recall that

$$|P_E x| := \max_{e \in \mathbb{S}^{d-1}} \langle P_E x, e \rangle.$$

Let $x_0 \in G$. The stability of minimizers implies that, for every $e \in S^{d-1}$, the test function

$$\psi(x) := \varphi(x) + \varepsilon \langle P_E x, e \rangle$$

touches u from below at x_0, provided $0 < \varepsilon \ll 1$ is taken small enough. Once again we resort to the equation for u to obtain

$$|M x_0 + \varepsilon e|^{\theta} F(M) \leq 0. \tag{5.19}$$

Now, we observe that

$$\sup_{e \in \mathbb{S}^{d-1}} |M x_0 + \varepsilon e| \geq \varepsilon.$$

Therefore, there exists $e^* \in \mathbb{S}^{d-1}$ such that $|Mx_0 + \varepsilon e^*| > \varepsilon/2$. By taking the supremum on both sides of (5.19), we obtain

$$\left[\frac{\varepsilon}{2}\right]^\theta F(M) \leq 0,$$

which in turn implies $F(M) \leq 0$.

Step 4. Finally, let x_0 be such that $P_e x_0 \neq 0$ and consider the unit vector v given by

$$v := \frac{P_E x_0}{|P_E x_0|}.$$

We notice that $|P_E x|$ is twice differentiable at x_0. Moreover

$$DP_E x_0 = v$$

and

$$D^2 P_E x_0 = I - v \otimes v.$$

The equation satisfied by u leads to

$$|Mx_0 + \varepsilon v|^\theta F (M + \varepsilon (\mathrm{Id} - v \otimes v)) \leq 0. \qquad (5.20)$$

In the following, let $(\sigma_i)_{i=1}^d$ be the basis formed by the eigenvectors of M. Write x_0 in this basis as $x_0 = \sum_{i=1}^d a_i \sigma_i$. Hence,

$$Mx_0 = \lambda_1 a_1 \sigma_1 + \cdots + \lambda_k a_k \sigma_k + \lambda_{k+1} a_{k+1} \sigma_{k+1} + \cdots + \lambda_d a_d \sigma_d,$$

with $\lambda_i > 0$ for $i = 1, 2, \ldots, k$. We conclude

$$|Mx_0 + \varepsilon v| \geq \langle Mx_0 + \varepsilon v, v \rangle$$

$$= \frac{1}{|P_E x_0|} \left\langle \sum_{i=1}^d \lambda_i a_i \sigma_i, \sum_{l=1}^k a_l \sigma_l \right\rangle + \varepsilon$$

$$= \frac{1}{|P_E x_0|} \sum_{i=1}^k \lambda_i a_i^2 + \varepsilon$$

$$\geq \varepsilon.$$

By multiplying (5.20) by $|Mx_0 + \varepsilon v|^{-\theta}$ and noticing that $\varepsilon (\mathrm{Id} - v \otimes v) \geq 0$, we are led to

$$F(M) \leq F (M + \varepsilon (\mathrm{Id} - v \otimes v)) \leq 0,$$

which completes the argument.   □

5.2.2 Compactness of Solutions

The former result unlocks an important connection between equations of the form

$$|Du + q|^\theta F(D^2 u) = f \quad \text{in} \quad B_1 \tag{5.21}$$

and the homogeneous uniformly elliptic problem $F = 0$. The underlying strategy is to relate (5.17) first with its homogeneous counterpart. Then Proposition 5.12 completes the argument. This connection follows from an approximation lemma. The first step is a compactness result.

Before proceeding, we comment on the importance of (5.21). Since we are interested in the Hölder-continuity of the gradient of solutions, the geometric argument in use involves an affine function. In fact, we have to control the oscillation of $u - \ell$, where ℓ is affine. When examining the equation solved by this difference, we obtain precisely (5.21), where q stands for the linear part of ℓ. By *compactness* we understand a universal Hölder estimate for u, together with a modulus of continuity, independent of $q \in \mathbb{R}^d$.

Imbert and Silvestre (2013) tackled this problem for the first time. To remove the dependence on q from the estimate, they split the Euclidean space into two subsets, namely: B_{C_0} and $\mathbb{R}^d \setminus B_{C_0}$, where $C_0 > 0$ is a universal constant to be determined endogenously, within the argument. If $q \in \mathbb{R}^d \setminus B_{C_0}$, then solutions to (5.21) are Lipschitz-continuous. Conversely, if $q \in B_{C_0}$, the authors resort to an argument in Imbert (2011) and prove that solutions are, in fact, Hölder-continuous; see Imbert and Silvestre (2013, Lemma 5). In what follows, we adopt the same strategy. A fundamental ingredient at this level is the maximum principle; we state it next.

Proposition 5.13 (Maximum principle) *Let $G \colon B_1 \times \mathbb{R}^d \times S(d) \to \mathbb{R}$ be a degenerate elliptic operator. Let $u \in C(B_1)$ be a viscosity solution to*

$$G\big(x, Du, D^2 u\big) = 0 \quad \text{in} \quad B_1.$$

Suppose $U \subset B_1$ and define $v \colon U \times U \to \mathbb{R}$ as

$$v(x, y) := u(x) - u(y).$$

Let $\psi \in C^2(U \times U)$ and suppose $(\overline{x}, \overline{y}) \in U \times U$ is a local maximum of $v - \psi$. Then, for each $\varepsilon > 0$ there exist $X, Y \in S(d)$ such that

$$G(\overline{x}, D_x \psi(\overline{x}, \overline{y}), X) \le 0 \le G(\overline{y}, -D_y \psi(\overline{x}, \overline{y}), Y).$$

In addition, the matrix inequality

$$-\left(\frac{1}{\varepsilon} + \|A\|\right) I \le \begin{pmatrix} X & 0 \\ 0 & -Y \end{pmatrix} \le A + \varepsilon A^2$$

holds, where $A := D^2 \psi(\overline{x}, \overline{y})$.

For a proof of Proposition 5.13 we refer the reader to Crandall et al. (1992). In the following, we apply the maximum principle in the case where $q \ge C_0$ and obtain Lipschitz-continuity of the solutions; this is the context of the next proposition.

Proposition 5.14 (Lipschitz-continuity, $|q| \ge C_0$)   *Let $u \in C(B_1)$ be a viscosity solution to*

$$|Du + q|^\theta F(D^2 u) = f \qquad in \qquad B_1,$$

where F is a (λ, Λ)-elliptic operator, $f \in L^\infty(B_1) \cap C(B_1)$ and $\theta > 0$. Let $q \in \mathbb{R}^d \setminus B_{C_0}$ be an arbitrary vector, where $C_0 > 0$ is yet to be determined. There exists $C > 0$ such that u is locally Lipschitz-continuous in B_1 and

$$|u(x) - u(y)| \le C|x - y|,$$

For every $x, y \in B_{1/2}$. The constant C depends on the data of the problem, but do not depend on the vector q.

Proof   We start by fixing $0 < r \ll 1$, to be determined further, and consider

$$L := \sup_{x,y \in B_r} \left(u(x) - u(y) - L_1 \omega(|x - y|) - L_2 \left(|x - x_0|^2 + |y - x_0|^2\right) \right),$$

defined for every $x_0 \in B_{r/2}$ where $\omega : \mathbb{R} \to \mathbb{R}$ is given by $\omega(t) := t - \frac{1}{2}t^2$. Our goal is to prove that $L \le 0$ for some $L_1, L_2 > 0$; from this fact, the result follows.

The proof is based on a contradiction argument. We suppose that for every $L_1 > 0$ and $L_2 > 0$, there is $x_0 \in B_{r/2}$ for which $L > 0$. In what follows, we split the proof into four steps.

Step 1. Let $\psi, \phi : \overline{B_r} \times \overline{B_r} \to \mathbb{R}$, be given by

$$\psi(x, y) := L_1 \omega(|x - y|) + L_2 \left(|x - x_0|^2 + |y - x_0|^2\right)$$

and

$$\phi(x, y) := u(x) - u(y) - \psi(x, y).$$

In addition, let $(\overline{x}, \overline{y}) \in \overline{B_r} \times \overline{B_r}$ be a maximum point for ϕ. Hence,

$$0 < L = \phi(\overline{x}, \overline{y}).$$

Because u is a normalized solution, we get

$$\psi(\bar{x}, \bar{y}) < u(\bar{x}) - u(\bar{y}) \le \mathrm{osc}_{B_1} u \le 2.$$

As a consequence,

$$L_1 \omega(|\bar{x} - \bar{y}|) + L_2\big(|\bar{x} - x_0|^2 + |\bar{y} - x_0|^2\big) \le 2.$$

At this point, we choose $L_2 > 0$. This choice is made to guarantee that $\bar{x}$ and $\bar{y}$ are in the interior of B_r; this condition is technical, though necessary. Set

$$L_2 := \left(\frac{4\sqrt{2}}{r}\right)^2$$

and observe that

$$|\bar{x} - x_0| \le \frac{r}{4} \quad \text{and} \quad |\bar{y} - x_0| \le \frac{r}{4}.$$

Then $\bar{x}, \bar{y} \in B_r$. Lastly, we note $\bar{x} \ne \bar{y}$; in fact, were $\bar{x} = \bar{y}$, immediately $L \le 0$.

Step 2. We proceed through an application of the maximum principle, as stated in the Proposition 5.13. Compute $D_x \psi$ and $D_y \psi$ and evaluate those quantities at $(\bar{x}, \bar{y})$. We get

$$D_x \psi(\bar{x}, \bar{y}) = L_1 \omega'(|\bar{x} - \bar{y}|)|\bar{x} - \bar{y}|^{-1}(\bar{x} - \bar{y}) + 2L_2(\bar{x} - x_0),$$

and

$$-D_y \psi(\bar{x}, \bar{y}) = L_1 \omega'(|\bar{x} - \bar{y}|)|\bar{x} - \bar{y}|^{-1}(\bar{x} - \bar{y}) - 2L_2(\bar{x} - x_0).$$

To simplify notation, introduce

$$q_{\bar{x}} := D_x \psi(\bar{x}, \bar{y}) \quad \text{and} \quad q_{\bar{y}} := D_y \psi(\bar{x}, \bar{y}).$$

For every $\iota > 0$, the maximum principle in Proposition 5.13 yields matrices $X, Y \in S(d)$ such that

$$|q_{\bar{x}} + q|^\theta F(X) - f(\bar{x}) \le 0 \le |q_{\bar{y}} + q|^\theta F(Y) - f(\bar{y}). \tag{5.22}$$

Moreover,

$$\begin{pmatrix} X & 0 \\ 0 & -Y \end{pmatrix} \le \begin{pmatrix} Z & -Z \\ -Z & Z \end{pmatrix} + (2L_2 + \iota)I, \tag{5.23}$$

and the matrix Z has the form

$$Z := L_1 \omega''(|\bar{x} - \bar{y}|)\frac{(\bar{x} - \bar{y}) \otimes (\bar{x} - \bar{y})}{|\bar{x} - \bar{y}|^2} + L_1 \frac{\omega'(|\bar{x} - \bar{y}|)}{|\bar{x} - \bar{y}|}\left(I - \frac{(\bar{x} - \bar{y}) \otimes (\bar{x} - \bar{y})}{|\bar{x} - \bar{y}|^2}\right).$$

Step 3. We recall that the extremal operators $\mathcal{P}^{\pm}_{\lambda,\Lambda}(M)$ depend on the spectrum of M. To explore this feature, we apply the matrix inequality (5.23) to appropriate vectors and recover information on the eigenvalues of $X - Y$. Take $v \in \mathbb{S}^{d-1}$ and consider first $(v, v) \in \mathbb{R}^{2d}$; we get

$$\langle (X - Y)v, v \rangle \le (4L_2 + 2\iota),$$

Because v is an arbitrary direction in $\mathbb{S}^{d-1}$, we conclude the eigenvalues of $X - Y$ are smaller than $4L_2 + 2\varepsilon\eta$. In addition, we apply (5.23) to vectors $(\bar{v}, -\bar{v}) \in \mathbb{R}^{2d}$, with

$$\bar{v} := \frac{\bar{x} - \bar{y}}{|\bar{x} - \bar{y}|};$$

we then get

$$\left\langle (X - Y)\frac{\bar{x} - \bar{y}}{|\bar{x} - \bar{y}|}, \frac{\bar{x} - \bar{y}}{|\bar{x} - \bar{y}|} \right\rangle \le (4L_2 + 2\iota)\left|\frac{\bar{x} - \bar{y}}{|\bar{x} - \bar{y}|}\right|^2$$

$$+ 4L_1\omega''(|\bar{x} - \bar{y}|) \qquad (5.24)$$

$$= 4L_2 + 2\iota - 4L_1.$$

Although (5.24) does not provide information on the entire spectrum of $X - Y$, it ensures that at least one of its eigenvalues is smaller than $-4L_1 + 4L_2 + 2\iota$. For $L_1 \gg 1$, we have

$$-4L_1 + 4L_2 + 2\iota < 0.$$

Now, we relate the former computation with the uniform ellipticity of F, through the extremal operator $\mathcal{P}^-_{\lambda,\Lambda}$. In fact,

$$\mathcal{P}^-_{\lambda,\Lambda}(X - Y) \ge 4\lambda L_1 - (\lambda + (d - 1)\Lambda)(4L_2 + 2\iota);$$

together with (5.22) it produces

$$4\lambda L_1 \le (\lambda + (d - 1)\Lambda)(4L_2 + 2\iota) + \frac{f(\bar{x})}{|q_{\bar{x}} + q|^\theta} - \frac{f(\bar{y})}{|q_{\bar{y}} + q|^\theta}. \qquad (5.25)$$

Step 4. Here, the assumption that $|q| \ge C_0$, for $C_0 > 0$ to be selected, is used. We continue by controlling the norm of $q_{\bar{x}}$; from its definition, we obtain

$$|q_{\bar{x}}| \le L_1|w'(|\bar{x} - \bar{y}|)| + 2L_2 \le cL_1, \qquad (5.26)$$

for some universal constant $c > 0$. Set $C_0 := \pi c L_1$, for L_1 yet to be selected. Because $|q_{\bar{x}}| < cL_1$ and $|q| > \pi c L_1$ we get

$$|q + q_{\bar{x}}| \ge C_0 - \frac{C_0}{\pi} = \frac{\pi - 1}{\pi}C_0;$$

similar reasoning yields

$$|q + q_{\bar{y}}| \geq C_0 - \frac{C_0}{\pi} = \frac{\pi - 1}{\pi} C_0.$$

The former inequalities give

$$\frac{f(\bar{x})}{|q + q_{\bar{x}}|^\theta} \leq \frac{\|f\|_{L^\infty(B_1)}}{\left|\frac{(\pi - 1)C_0}{\pi}\right|^\theta} \leq \|f\|_{L^\infty(B_1)} \tag{5.27}$$

and

$$\frac{f(\bar{y})}{|q + q_{\bar{y}}|^\theta} \leq \frac{\|f\|_{L^\infty(B_1)}}{\left|\frac{(\pi - 1)C_0}{\pi}\right|^\theta} \leq \|f\|_{L^\infty(B_1)}. \tag{5.28}$$

Inequalities (5.27) and (5.28), combined with (5.25), produce

$$4\lambda L_1 \leq (\lambda + (d - 1)\Lambda)(4L_2 + 2\iota) + 2\|f\|_{L^\infty(B_1)}. \tag{5.29}$$

If we choose $L_1 = L_1(\lambda, \Lambda, d, L_2, r) \gg 1$ large enough in (5.29), we reach a contradiction. Such a contradiction implies $L \leq 0$ and finishes the proof. $\quad\square$

To complete the analysis of compactness for the solutions to (5.21), it remains for us to consider the case $|q| < C_0$, for the constant $C_0 > 0$ determined in the proof of Proposition 5.14. To that end, we recall a result first presented in Imbert and Silvestre (2016).

In fact, Imbert and Silvestre (2016) address viscosity solutions to equations *holding only where the gradient of the solutions is large*. For example, let $u \in C(B_1)$ be a solution to

$$\mathcal{P}_{\lambda,\Lambda}^-(D^2 u) + \gamma |Du| \leq C_0 \quad \text{in} \quad B_1 \cap \{|Du| > 1\},$$

where γ and C_0 are given constants. The findings of Imbert and Silvestre (2016) comprise a Harnack inequality for u, together with a *local* Hölder-continuity result. Hence, although the region where the equation holds is unknown a priori, $u \in C_{\text{loc}}^\alpha(B_1)$, for some $\alpha \in (0, 1)$. The case of measurable ingredients, namely $\gamma, C_0 \in L^d(B_1)$, is the subject of Mooney (2015).

When studying the compactness of solutions in the case $|q| < C_0$, we use the results in Imbert and Silvestre (2016). For the sake of completeness, we recall it next, adapted to our setting.

Lemma 5.15 *Let* $u \in C(B_1)$ *be a viscosity solution to*

$$G(x, Du, D^2 u) = 0 \quad \text{in} \quad B_1,$$

where $G \in C(B_1 \times \mathbb{R}^d \times S(d))$. Suppose there exists $A_0 > 0$ such that $G(x, p, M)$ is (λ, Λ)-elliptic whenever $|p| \geq A_0$. Then, $u \in C_{loc}^{\beta}(B_1)$ and there exists $C > 0$, depending on d, λ, Λ and A_0, such that

$$\|u\|_{C^{\beta}(B_{1/2})} \leq C \|u\|_{L^{\infty}(B_1)}.$$

See Imbert and Silvestre (2016, Theorem 1.1). Now, we suppose $|q| < C_0$ and frame (5.21) in the context of Lemma 5.15. It produces Hölder-continuity for the solutions to that equation and, combined with Proposition 5.14, leads to the compactness of the solutions.

Proposition 5.16 (Hölder-continuity, $|q| < C_0$) *Let $u \in C(B_1)$ be a viscosity solution to*

$$|Du + q|^{\theta} F(D^2 u) = f \qquad in \qquad B_1,$$

where F is a (λ, Λ)-elliptic operator, $f \in L^{\infty}(B_1) \cap C(B_1)$ and $\theta > 0$. Let $q \in B_{C_0}$ be an arbitrary vector, where $C_0 > 0$ is the constant determined in Proposition 5.14. There exist $\beta \in (0, 1)$ and $C > 0$ such that $u \in C_{loc}^{\beta}(B_1)$ and

$$\|u\|_{C^{\beta}(B_{1/2})} \leq C \left(\|u\|_{L^{\infty}(B_1)} + \|f\|_{L^{\infty}(B_1)}^{\frac{1}{1+\theta}} \right).$$

The constants β and C depend on the data of the problem and C_0, but do not depend on the vector q.

Proof Let $C_0 > 0$ be given as in the proof of Proposition 5.14. Define the operator

$$G(x, p, M) := |q + p|^{\theta} F(M) - f(x).$$

It follows that $G(x, p, M)$ is uniformly elliptic provided $|p| > \pi C_0$. A straightforward application of Lemma 5.15 implies the Hölder-continuity of the solutions.   □

Compare the proof of Proposition 5.16 with the analysis in Imbert and Silvestre (2016, p. 1323). c.f. Imbert and Silvestre (2013, Proof of Lemma 5).

5.2.3 Optimal Regularity of Solutions

In this section we complete the proof of Theorem 5.11. We start by combining the Cancellation Lemma in Proposition 5.12 with standard facts in the theory of viscosity solutions to produce an approximation lemma.

Proposition 5.17 *Let $u \in C(B_1)$ be a viscosity solution to*

$$|Du + q|^\theta F(D^2 u) = f \qquad in \qquad B_1,$$

where F is a (λ, Λ)-elliptic operator, $f \in L^\infty(B_1) \cap C(B_1)$, $\theta > 0$ is fixed, and $q \in \mathbb{R}^d$. Given $\delta > 0$, one can find $h \in C_{loc}^{1,\alpha_0}(B_{9/10})$, satisfying

$$\|u - h\|_{L^\infty(B_{9/10})} \leq \delta,$$

where $\alpha_0 \in (0,1)$ is the exponent governing the C^{1,α_0}-regularity associated with $F = 0$. Moreover,

$$\|h\|_{C^{1,\beta}(B_{9/10})} \leq C \|h\|_{L^\infty(B_1)}, \tag{5.30}$$

for some $C > 0$ depending only on d, λ and Λ.

Proof We establish the result through a contradiction argument. Suppose the statement of the proposition is false. It amounts to the existence of sequences of functions $(u_n)_{n \in \mathbb{N}} \subset C(B_1)$ and $(f_n)_{n \in \mathbb{N}} \subset L^\infty(B_1) \cap C(B_1)$, a sequence of vectors $(q_n)_{n \in \mathbb{N}} \subset \mathbb{R}^d$ and a sequence of (λ, Λ)-elliptic operators $(F_n)_{n \in \mathbb{N}}$ such that

$$|Du_n + q_n|^\theta F_n(D^2 u_n) = f_n \qquad in \qquad B_1, \tag{5.31}$$

$$\|f_n\|_{L^\infty(B_1)} \leq \frac{1}{n},$$

but

$$\|u_n - h\|_{L^\infty(B_1)} > \delta_0,$$

for every $n \in \mathbb{N}$, every $h \in C^{1,\alpha_0}(B_1)$, and some $\delta_0 > 0$.

From Propositions 5.14 and 5.16, we infer that $(u_n)_{n \in \mathbb{N}}$ is equibounded in some C^β space. Hence, through a subsequence if necessary, $u_n \to u_\infty$ locally uniformly, for some $u_\infty \in C_{loc}^{\beta/2}(B_1)$. In addition, (λ, Λ)-ellipticity implies that $(F_n)_{n \in \mathbb{N}}$ is uniformly Lipschitz-continuous. As a consequence, there exists a (λ, Λ)-elliptic operator F_∞ such that $F_n \to F_\infty$, locally uniformly. At this point we distinguish two cases.

Case 1. Suppose $(q_n)_{n \in \mathbb{N}}$ admits a convergent subsequence, still denoted with $(q_n)_{n \in \mathbb{N}}$, with limit $q_\infty \in \mathbb{R}^d$. The stability of viscosity solutions implies that

$$|Du_\infty + q_\infty|^\theta F_\infty(D^2 u_\infty) = 0 \qquad in \qquad B_1.$$

Proposition 5.12, in turn, leads to $F_\infty(D^2 u_\infty) = 0$. We immediately infer that $u_\infty \in C_{loc}^{1,\alpha_0}(B_{9/10})$, with estimates.

Case 2. Suppose $(q_n)_{n\in\mathbb{N}}$ does not admit a convergent subsequence. It means $(q_n)_{n\in\mathbb{N}}$ in unbounded. Consider the sequence of directions $e_n := q_n/|q_n|$, for every $n \in \mathbb{N}$. Thus, (5.31) becomes

$$|a_n Du_n + e_n|^\theta \, F_n(D^2 u_n) = \frac{f_n}{|q_n|^\theta} \qquad \text{in} \qquad B_1,$$

where

$$a_n := \frac{1}{|q_n|^\theta} \longrightarrow 0.$$

It is clear that $e_n \to e_\infty$ through some subsequence, if necessary. In the limit, we get

$$|e_\infty + 0Du_\infty|^\theta \, F(D^2 u_\infty) = 0,$$

in the unit ball. Proposition 5.12 also produces $F_\infty(D^2 u_\infty) = 0$ in this case. Once again, the regularity theory for uniformly elliptic homogeneous equations ensures $C^{1,\alpha_0}_{\text{loc}}(B_1)$.

In both cases, we can take $h := u_\infty$, which produces a contradiction and completes the proof.   □

The regularity of the approximation function plays an important role in the sequel. The Taylor expansion for C^{1,α_0}-regular functions leads to

$$\sup_{x\in B_r(x_0)} |h(x) - h(x_0) - Dh(x_0) \cdot (x - x_0)| \leq Cr^{1+\alpha_0},$$

for every $r > 0$ and $x_0 \in B_{9/10}$ such that $B_r(x_0) \Subset B_{9/10}$, where $C > 0$ is the universal constant in (5.30). We continue with an elementary oscillation control for the solutions to (5.21). At this point we fix $\alpha \in (0,\alpha_0)$ satisfying

$$\alpha \leq \frac{1}{1+\theta}.$$

Lastly, given α, we set the universal constant

$$\rho := \left(\frac{1}{2C}\right)^{\frac{1}{\alpha_0-\alpha}}, \tag{5.32}$$

where $C > 0$ is the constant in (5.30). The constants α and ρ remain unchanged throughout the remainder of this section

Proposition 5.18 (Oscillation control) *Let* $u \in C(B_1)$ *be a viscosity solution to*

$$|Du + q|^\theta F(D^2 u) = f \qquad \text{in} \qquad B_1,$$

where F is a (λ, Λ)-elliptic operator, $f \in L^{\infty}(B_1) \cap C(B_1)$, the exponent $\theta > 0$ is fixed, and $q \in \mathbb{R}^d$ is arbitrary. Then there exist an affine function $\ell(x) := a + b \cdot x$ and a universal constant $C > 0$, such that

$$|a| + |b| \leq C,$$

and

$$\sup_{x \in B_\rho} |u(x) - \ell(x)| \leq \rho^{1+\alpha}.$$

Proof Set $a := h(0)$ and $b := Dh(0)$, where h is the approximating function whose existence is ensured in Proposition 5.17. The triangle inequality yields

$$\sup_{x \in B_\rho} |u(x) - h(0) - Dh(0) \cdot x| \leq \sup_{x \in B_\rho} |u(x) - h(x)|$$

$$+ \sup_{x \in B_\rho} |h(x) - h(0) - Dh(0) \cdot x|$$

$$\leq \delta + C\rho^{1+\beta}.$$

Fix

$$\delta := \frac{\rho^{1+\alpha}}{2};$$

notice this (universal) choice sets the smallness regime in Proposition 5.17. The choice of ρ in (5.32) implies

$$\sup_{x \in B_\rho} |u(x) - \ell(x)| \leq \rho^{1+\alpha}.$$

Also, the estimate for h gives $|a| + |b| \leq C$ and completes the proof. $\square$

The next proposition concerns the oscillation at discrete scales ρ^n, for $n \in \mathbb{N}$.

Proposition 5.19 (Oscillation control at discrete scales)   *Let $u \in C(B_1)$ be a viscosity solution to*

$$|Du + q|^\theta F(D^2u) = f \qquad in \;\; B_1,$$

where F is a (λ, Λ)-elliptic operator, $f \in L^{\infty}(B_1) \cap C(B_1)$, the exponent $\theta > 0$ is fixed, and $q \in \mathbb{R}^d$ is arbitrary. Then there exists a sequence of affine function $(\ell_n)_{n\in\mathbb{N}}$, given by

$$\ell_n(x) := a_n + b_n \cdot x,$$

satisfying

$$\sup_{x \in B_{\rho^n}} |u(x) - \ell_n(x)| \leq \rho^{(1+\alpha)n} \tag{5.33}$$

and

$$|a_n - a_{n-1}| + |b_n - b_{n-1}|\rho^{n-1} \leq C\rho^{(1+\alpha)(n-1)}, \tag{5.34}$$

where $C > 0$ is a universal constant.

Proof The proof follows from an induction argument, which we split into three main steps, for ease of clarity.

Step 1. We start by considering $n = 1$; in fact, set $\ell_0(x) \equiv 0$ and $\ell_1(x) := h(0) + Dh(0) \cdot x$, where h is the function in the proof of Proposition 5.18. The case $n = 1$ follows at once from Proposition 5.18.

Step 2. Suppose the case $n = k$ has already been verified; we examine next the case $n = k + 1$. Consider the auxiliary function

$$v_k(x) := \frac{u(\rho^k x) - \ell_k(\rho^k x)}{\rho^{k(1+\alpha)}}.$$

The function v_k solves

$$|Dv_k + b_k|^\theta F_k(D^2 v_k) = f_k,$$

where

$$F_k(M) := \rho^{(1-\alpha)k} F(\rho^{(\alpha-1)k} M) \quad \text{and} \quad f_k(x) := \rho^{k(1-\alpha-\alpha\theta)} f(\rho^k x).$$

It is clear that F_k inherits the (λ, Λ)-ellipticity of F. Also, $f_k \in C(B_1) \cap L^\infty(B_1)$, since

$$\alpha \leq \frac{1}{1+\theta}.$$

Therefore, v_k meets the conditions of Proposition 5.17. As a consequence, there exists $\overline{h} \in C^{1,\alpha_0}_{\text{loc}}(B_1)$ such that

$$\sup_{x \in B_\rho} |v_k(x) - \overline{h}(x)| \leq \delta.$$

Proposition 5.18 is also available for v_k, yielding

$$\sup_{x \in B_\rho} |v_k(x) - \overline{h}(0) - D\overline{h}(0) \cdot x| \leq \rho. \tag{5.35}$$

The definition of v_k builds upon (5.35) to produce

$$\sup_{x \in B_\rho} \left| \frac{u\left(\rho^k x\right) - a_k - b_x \cdot \left(\rho^k x\right) - \rho^{k(1+\alpha)}\overline{h}(0) - \rho^{k\alpha} D\overline{h}(0) \cdot \left(\rho^k x\right)}{\rho^{k(1+\alpha)}} \right| \le \rho.$$

(5.36)

Define $\ell_{k+1}(x)$ as

$$\ell_{k+1}(x) := a_{k+1} + b_{k+1} \cdot x,$$

where

$$a_{k+1} := a_k + \rho^{k(1+\alpha)}\overline{h}(0) \quad \text{and} \quad b_{k+1} := b_k + \rho^{k\alpha} D\overline{h}(0). \quad (5.37)$$

Step 3. From (5.36) we infer that

$$\sup_{x \in B_{\rho^{k+1}}} |u(x) - \ell_{k+1}(x)| \le \rho^{n+1}.$$

By noticing that $\overline{h}$ has the same estimates as h in Proposition 5.18 and resorting to (5.37), we get

$$|a_{k+1} - a_k| = \rho^{k(1+\alpha)}\overline{h}(0) \le C\rho^{k(1+\alpha)}$$

and

$$|b_{k+1} - b_k| \rho^k = \rho^{k(1+\alpha)}D\overline{h}(0) \le C\rho^{k(1+\alpha)},$$

which completes the proof. □

Next, we detail the proof of Theorem 5.11. It relies on two main ingredients: an oscillation control on a continuous scale and a convergence analysis.

Proof of Theorem 5.11   The proof amounts to two main steps. First, we notice that $(\ell_n)_{n \in \mathbb{N}}$ is a convergent consequence.

Step 1. From (5.34) we learn that $(a_n)_{n \in \mathbb{N}}$ and $(b_n)_{n \in \mathbb{N}}$ are Cauchy sequences. Therefore there exist $a_\infty \in \mathbb{R}$ and $b_\infty \in \mathbb{R}^d$ such that $a_n \to a_\infty$ and $b_n \to b_\infty$. Moreover, (5.33) yields $a_\infty \equiv u(0)$ and $b_\infty \equiv Du(0)$, with

$$|u(0) - a_n| < C\rho^{(n-1)(1}$$

Step 2. Next, we compute the oscillation of u, subtracted $(u(0) - Du(0) \cdot x)$, in balls with arbitrarily small radius r. Let $m \in \mathbb{N}$ be such that $\rho^{m+1} \le r < \rho^m$; from Proposition 5.19 we obtain

$$\sup_{x \in B_r} |u(x) - u(0) - Du(0) \cdot x| \leq \sup_{x \in B_{\rho^m}} |u(x) - u(0) - Du(0) \cdot x|$$

$$\leq \sup_{x \in B_{\rho^m}} |u(x) - \ell_m(x)|$$

$$+ \sup_{x \in B_{\rho^m}} |\ell_m(x) - u(0) - Du(0) \cdot x|$$

$$\leq C\rho^{m(1+\alpha)}$$

$$= \frac{C}{\rho^{1+\alpha}} \rho^{(m+1)(1+\alpha)},$$

which ends the proof.   □

We finish this chapter with a few comments on degenerate models, both in the variational as well as in the nonvariational setting.

5.3 Further Remarks on Degenerate Diffusions

One important variant of the problem in Section 5.1 concerns the parabolic setting. Namely, the p-caloric equation

$$u_t - \text{div}\left(|Du|^{p-2}Du\right) = f \quad \text{in} \quad B_1 \times (-1,0] =: Q_1, \tag{5.38}$$

where $p > 1$ and $f \in L^r(-1,0; L^q(B_1))$, for some $1 < r,q \leq \infty$. When it comes to the regularity theory of (5.38), and related degenerate or singular models, it is paramount to mention the fundamental idea of *intrinsic scaling*; see, for instance, DiBenedetto (1986) and the comprehensive monograph by Urbano (2008). This notion translates the structure of the PDE into a relation for the temporal and spatial scales. In this context, one examines the oscillation of the solutions in a family of cylinders whose geometry reflects the equation. The breakthrough, as one can read in the introduction to Urbano (2008), is that "the equation behaves, in its own geometry, like the heat equation."

The regularity of the solutions to (5.38), and related problems, is the subject of important contributions; see DiBenedetto (1982, 1983, 1987); DiBenedetto and Friedman (1985); DiBenedetto et al. (2004, 2007); Acerbi and Mingione (2005, 2007); Duzaar and Mingione (2005); and Kuusi and Mingione (2013, 2014a,b), to name only a few. However, an instance of particular interest is the *explicit characterization* of the Hölder-modulus of continuity for the solutions. Such a characterization sits at the intersection of methods in the realm of intrinsic scaling and approximation techniques.

In fact, Teixeira and Urbano (2014) prove that solutions to (5.38) are locally of class C^α in space and $C^{\alpha/2\alpha+(1-\alpha)p}$ in time. Here,

$$\alpha := \frac{(pq - d)r - pq}{q[(p - 1)r - (p - 2)]}.$$

Another issue of interest is the L^p-viscosity theory associated with

$$|Du|^\theta F(D^2 u) = f \quad \text{in} \quad B_1, \tag{5.39}$$

where the continuity assumption on the source term f is dropped, and we have $\|f\|_{L^\infty(B_1)} \leq C_0$. For $p \in \mathbb{R}^d$ and $M \in S(d)$, write

$$G(x, p, M) := |p|^\theta F(M) - f(x).$$

Then the operator driving (5.39) satisfies structural conditions of the form

$$-C_0 + |p|^\theta \mathcal{P}^-_{\lambda,\Lambda}(M) \leq G(x, p, M) \leq |p|^\theta \mathcal{P}^+_{\lambda,\Lambda}(M) + C_0, \tag{5.40}$$

and

$$\begin{aligned}
|p|^\theta \mathcal{P}^-_{\lambda,\Lambda}(M - N) + \left(|p|^\theta - |q|^\theta\right)\mathcal{P}^-(N) \\
\leq G(x, p, M) - G(x, q, N) \\
\leq |p|^\theta \mathcal{P}^+_{\lambda,\Lambda}(M - N) + \left(|p|^\theta - |q|^\theta\right)\mathcal{P}^+(N),
\end{aligned} \tag{5.41}$$

for $p, q \in \mathbb{R}^d$ and $M, N \in S(d)$. In the purely degenerate case $\theta > 0$, the Definition 1.11 of L^p-viscosity solution is suitable. In the singular case we must consider an alternative definition, in the spirit of Evans and Spruck (1991, Definition 2.1); see also Birindelli and Demengel (2004). The remaining question is whether or not a program inspired by the results in Caffarelli et al. (1996) is available under structural conditions in line with (5.40) and (5.41).

We conclude these remarks with a problem bringing together the variational and nonvariational approaches. Consider the *thermistor problem*

$$\begin{cases}
-\operatorname{div}\left(|Du|^{\sigma(\theta(x))-2}\right) = f & \text{in} \quad B_1 \\
-\Delta\theta = \lambda\left(\theta(x)\right)|Du|^{\sigma(\theta(x))} & \text{in} \quad B_1.
\end{cases} \tag{5.42}$$

Introduced in Zhikov (2008), this system models the equilibrium distribution of the electrical potential u and the temperature θ in a solid, where the functions $f: B_1 \to \mathbb{R}$ and $\sigma, \lambda: \mathbb{R} \to \mathbb{R}$ are given. The first equation in (5.42) is of p-Laplacian type, with an exponent depending on θ through a given function $\sigma: \mathbb{R} \to \mathbb{R}$. The second equation in the system is a semilinear problem depending on u through its gradient.

Pimentel and Urbano (2021) examine (5.42) developing an existence theory in arbitrary dimensions $d \geq 3$ and studying the regularity of solutions. Approximation methods play an important role in Pimentel and Urbano (2021), as they relate the temperature θ in the thermistor problem with harmonic functions. They allow the authors to prove $\theta \in C_{\mathrm{loc}}^{1,\,\mathrm{Log\text{-}Lip}}(B_1)$.

Bibliographic Notes

For a modern account of the p-Laplace and the ∞-Laplace operators, we suggest the monographs of Lindqvist (2006) and (2016), respectively. Evans (2007a) discusses the 1-Laplacian, the ∞-Laplacian, and a game theoretic perspective of those models. As mentioned in the text, the regularity conjectures involving the p-Poisson equations have been proved in dimension $d = 2$ by Araújo et al. (2017) and Lindgren and Lindqvist (2017); see also the works of Araújo et al. (2018) and Iwaniec and Manfredi (1989).

The regularity theory for the p-Poisson equation has a counterpart in the study of problems in the calculus of variations; for a list of developments in this direction, including functionals with nonstandard growth, we mention the works of Mingione (2011), Acerbi and Mingione (2001a,b, 2002a,b, 2005), De Filippis and Mingione (2020, 2021), Duzaar and Mingione (2011), and Kuusi and Mingione (2012, 2013, 2014a,b). Equations holding only where the gradient is large are introduced and treated by Imbert and Silvestre (2016). An extension in the presence of measurable ingredients is discussed by Mooney (2015).

For the nonvariational problem modeled after the p-Laplace operator, we refer the reader to the works of Birindelli and Demengel (2004, 2006, 2007a,b, 2009, 2014) for results covering comparison principles, Liouville Theorems, the study of spectral properties of those operators, and Hölder-continuity of the gradient. When it comes to regularity of the solutions we also mention the works of Imbert and Silvestre (2013) and Araújo et al. (2015). See also the developments reported by De Filippis (2021b), for a more general class of operators. Finally, we mention the work of Araújo and Sirakov (2021) for an account of boundary regularity; c.f. Silvestre and Sirakov (2014).

For the analysis of the fully nonlinear model in the presence of variable exponents see Bronzi et al. (2020). For transmission problems related to this model we refer the reader to Huaroto et al. (2020), and the contributions of Colombo et al. (2021) and De Filippis (2021a).

References

Acerbi, E., and Mingione, G. 2001a. Regularity results for a class of functionals with non-standard growth. *Arch. Ration. Mech. Anal.*, **156**(2), 121–140.

Acerbi, E., and Mingione, G. 2001b. Regularity results for a class of quasiconvex functionals with nonstandard growth. *Ann. Scuola Norm. Sup. Pisa Cl. Sci. (4)*, **30**(2), 311–339.

Acerbi, E., and Mingione, G. 2002a. Regularity results for electrorheological fluids: the stationary case. *C. R. Math. Acad. Sci. Paris*, **334**(9), 817–822.

Acerbi, E., and Mingione, G. 2002b. Regularity results for stationary electrorheological fluids. *Arch. Ration. Mech. Anal.*, **164**(3), 213–259.

Acerbi, E., and Mingione, G. 2005. Gradient estimates for the $p(x)$-Laplacean system. *J. Reine Angew. Math.*, **584**, 117–148.

Acerbi, E., and Mingione, G. 2007. Gradient estimates for a class of parabolic systems. *Duke Math. J.*, **136**(2), 285–320.

Amaral, M. D., and Teixeira, E. V. 2015. Free transmission problems. *Comm. Math. Phys.*, **337**(3), 1465–1489.

Araújo, D., and Sirakov, B. 2021. Sharp boundary and global regularity for degenerate fully nonlinear elliptic equations. *arXiv preprint arXiv:2108.01150.*

Araújo, D., Ricarte, G., and Teixeira, E. 2015. Geometric gradient estimates for solutions to degenerate elliptic equations. *Calc. Var. Partial Differential Equations*, **53**(3–4), 605–625.

Araújo, D., Teixeira, E., and Urbano, J.-M. 2017. A proof of the $C^{p'}$-regularity conjecture in the plane. *Adv. Math.*, **316**, 541–553.

Araújo, D., Teixeira, E., and Urbano, J. M. 2018. Towards the $C^{p'}$-regularity conjecture in higher dimensions. *Int. Math. Res. Not. IMRN*, 6481–6495.

Araújo, D., Maia, A., and Urbano, J. M. 2020. Sharp regularity for the inhomogeneous porous medium equation. *J. Anal. Math.*, **140**(2), 395–407.

Armstrong, S., and Tran, H. 2015. Viscosity solutions of general viscous Hamilton–Jacobi equations. *Math. Ann.*, **361**(3–4), 647–687.

Armstrong, S., Silvestre, L., and Smart, C. 2012. Partial regularity of solutions of fully nonlinear, uniformly elliptic equations. *Comm. Pure Appl. Math.*, **65**(8), 1169–1184.

Bardi, M., and Capuzzo–Dolcetta, I. 1997. *Optimal Control and Viscosity Solutions of Hamilton–Jacobi-Bellman Equations.* Systems & Control: Foundations & Applications. With appendices by Maurizio Falcone and Pierpaolo Soravia. Birkhäuser Boston, Inc., Boston, MA.

Barles, G. 1994. *Solutions de viscosité des équations de Hamilton–Jacobi.* Mathématiques & Applications (Berlin) [Mathematics & Applications], vol. 17. Springer-Verlag, Paris.

Barron, E. N., Evans, L. C., and Jensen, R. 1984. Viscosity solutions of Isaacs' equations and differential games with Lipschitz controls. *J. Differential Equations,* **53**(2), 213–233.

Birindelli, I., and Demengel, F. 2004. Comparison principle and Liouville type results for singular fully nonlinear operators. *Ann. Fac. Sci. Toulouse Math. (6),* **13**(2), 261–287.

Birindelli, I., and Demengel, F. 2006. First eigenvalue and maximum principle for fully nonlinear singular operators. *Adv. Differential Equations,* **11**(1), 91–119.

Birindelli, I., and Demengel, F. 2007a. The Dirichlet problem for singular fully nonlinear operators. *Discrete Contin. Dyn. Syst. Proceedings of the 6th AIMS International Conference, suppl.,* pp. 110–121.

Birindelli, I., and Demengel, F. 2007b. Eigenvalue, maximum principle and regularity for fully nonlinear homogeneous operators. *Commun. Pure Appl. Anal.,* **6**(2), 335–366.

Birindelli, I., and Demengel, F. 2009. Eigenvalue and Dirichlet problem for fully-nonlinear operators in non-smooth domains. *J. Math. Anal. Appl.,* **352**(2), 822–835.

Birindelli, I., and Demengel, F. 2014. $C^{1,\beta}$ regularity for Dirichlet problems associated to fully nonlinear degenerate elliptic equations. *ESAIM Control Optim. Calc. Var.,* **20**(4), 1009–1024.

Birindelli, I., Galise, G., and Ishii, H. 2018. A family of degenerate elliptic operators: maximum principle and its consequences. *Ann. Inst. H. Poincaré Anal. Non Linéaire,* **35**(2), 417–441.

Bronzi, A., Pimentel, E., Rampasso, G., and Teixeira, E. 2020. Regularity of solutions to a class of variable-exponent fully nonlinear elliptic equations. *J. Funct. Anal.,* **279**(12), paper 108781, pages 31.

Cabré, X., and Caffarelli, L. 2003. Interior $C^{2,\alpha}$ regularity theory for a class of nonconvex fully nonlinear elliptic equations. *J. Math. Pures Appl. (9),* **82**(5), 573–612.

Caffarelli, L. 1988. Elliptic second order equations. *Rend. Sem. Mat. Fis. Milano,* **58**, 253–284.

Caffarelli, L. A. 1989. Interior a priori estimates for solutions of fully nonlinear equations. *Ann. of Math. (2),* **130**(1), 189–213.

Caffarelli, L. 1990. Interior $W^{2,p}$ estimates for solutions of the Monge–Ampère equation. *Ann. of Math. (2),* **131**(1), 135–150.

Caffarelli, L. 1991. Some regularity properties of solutions of Monge Ampère equation. *Comm. Pure Appl. Math.,* **44**(8–9), 965–969.

Caffarelli, L., and Cabré, X. 1995. *Fully Nonlinear Elliptic Equations.* American Mathematical Society Colloquium Publications, vol. 43. American Mathematical Society, Providence, RI.

Caffarelli, L., and Silvestre, L. 2010a. On the Evans–Krylov theorem. *Proc. Amer. Math. Soc.*, **138**(1), 263–265.

Caffarelli, L, and Silvestre, L. 2010b. Smooth approximations of solutions to nonconvex fully nonlinear elliptic equations. Pages 67–85 of: *Nonlinear Partial Differential Equations and Related Topics*. Amer. Math. Soc. Transl. Ser. 2, vol. 229. Amer. Math. Soc., Providence, RI.

Caffarelli, L., and Silvestre, L. 2011. The Evans–Krylov theorem for nonlocal fully nonlinear equations. *Ann. of Math. (2)*, **174**(2), 1163–1187.

Caffarelli, L. A., Crandall, M. G., Kocan, M., and Święch, A. 1996. On viscosity solutions of fully nonlinear equations with measurable ingredients. *Comm. Pure Appl. Math.*, **49**(4), 365–397.

Capuzzo Dolcetta, I., Leoni, F., and Porretta, A. 2010. Hölder estimates for degenerate elliptic equations with coercive Hamiltonians. *Trans. Amer. Math. Soc.*, **362**(9), 4511–4536.

Colombo, M., Kim, S., and Shahgholian, H. 2021. A transmission problem for (p,q)-Laplacian. *arXiv preprint arXiv:2106.07315*.

Crandall, M., and Lions, P.-L. 1983. Viscosity solutions of Hamilton–Jacobi equations. *Trans. Amer. Math. Soc.*, **277**(1), 1–42.

Crandall, M., Evans, L., and Lions, P.-L. 1984. Some properties of viscosity solutions of Hamilton–Jacobi equations. *Trans. Amer. Math. Soc.*, **282**(2), 487–502.

Crandall, M. G., Ishii, H., and Lions, P.-L. 1992. User's guide to viscosity solutions of second order partial differential equations. *Bull. Amer. Math. Soc. (N.S.)*, **27**(1), 1–67.

Crandall, M., Kocan, M., Soravia, P., and Święch, A. 1996. On the equivalence of various weak notions of solutions of elliptic PDEs with measurable ingredients. Pages 136–162 of: *Progress in Elliptic and Parabolic Partial Differential Equations (Capri, 1994)*. Pitman Res. Notes Math. Ser., vol. 350. Longman, Harlow.

Crandall, M., Kocan, M., and Święch, A. 2000. L^p-theory for fully nonlinear uniformly parabolic equations. *Comm. Partial Differential Equations*, **25**(11–12), 1997–2053.

De Filippis, C. 2021a. Fully nonlinear free transmission problems with nonhomogeneous degeneracies. *arXiv preprint arXiv:2103.12453*.

De Filippis, C. 2021b. Regularity for solutions of fully nonlinear elliptic equations with nonhomogeneous degeneracy. *Proc. Roy. Soc. Edinburgh Sect. A*, **151**(1), 110–132.

De Filippis, C., and Mingione, G. 2020. On the regularity of minima of non-autonomous functionals. *J. Geom. Anal.*, **30**(2), 1584–1626.

De Filippis, C., and Mingione, G. 2021. Interpolative gap bounds for nonautonomous integrals. *Anal. Math. Phys.*, **11**(3), paper 117, 39.

DiBenedetto, E. 1982. Continuity of weak solutions to certain singular parabolic equations. *Ann. Mat. Pura Appl. (4)*, **130**, 131–176.

DiBenedetto, E. 1983. $C^{1+\alpha}$ local regularity of weak solutions of degenerate elliptic equations. *Nonlinear Anal.*, **7**(8), 827–850.

DiBenedetto, E. 1986. On the local behaviour of solutions of degenerate parabolic equations with measurable coefficients. *Ann. Scuola Norm. Sup. Pisa Cl. Sci. (4)*, **13**(3), 487–535.

DiBenedetto, E. 1987. The flow of two immiscible fluids through a porous medium: regularity of the saturation. Pages 123–141 of: *Theory and Applications of Liquid Crystals (Minneapolis, Minn., 1985)*. IMA Vol. Math. Appl., vol. 5. Springer, New York.

DiBenedetto, E., and Friedman, A. 1985. Hölder estimates for nonlinear degenerate parabolic systems. *J. Reine Angew. Math.*, **357**, 1–22.

DiBenedetto, E., Urbano, J. M., and Vespri, V. 2004. Current issues on singular and degenerate evolution equations. Pages 169–286 of: *Evolutionary Equations. Vol. I*. Handb. Differ. Equ. North-Holland, Amsterdam.

DiBenedetto, E., Gianazza, U., Safonov, M., Urbano, J. M., and Vespri, V. 2007. Harnack's estimates: positivity and local behavior of degenerate and singular parabolic equations. *Bound. Value Probl.*, Art. ID 42548, 5.

Diehl, N., and Urbano, J. M. 2020. Sharp Hölder regularity for the inhomogeneous Trudinger's equation. *Nonlinearity*, **33**(12), 7054–7066.

Duzaar, F., and Mingione, G. 2005. Second order parabolic systems, optimal regularity, and singular sets of solutions. *Ann. Inst. H. Poincaré Anal. Non Linéaire*, **22**(6), 705–751.

Duzaar, F., and Mingione, G. 2011. Gradient estimates via non-linear potentials. *Amer. J. Math.*, **133**(4), 1093–1149.

Escauriaza, L. 1993. $W^{2,n}$ a priori estimates for solutions to fully nonlinear equations. *Indiana Univ. Math. J.*, **42**(2), 413–423.

Evans, L. C. 1982a. Classical solutions of fully nonlinear, convex, second-order elliptic equations. *Comm. Pure Appl. Math.*, **35**(3), 333–363.

Evans, Lawrence C. 1982b. A new proof of local $C^{1,\alpha}$ regularity for solutions of certain degenerate elliptic p.d.e. *J. Differential Equations*, **45**(3), 356–373.

Evans, L. 2007a. The 1-Laplacian, the ∞-Laplacian and differential games. Pages 245–254 of: *Perspectives in Nonlinear Partial Differential Equations*. Contemp. Math., vol. 446. Amer. Math. Soc., Providence, RI.

Evans, L. C. 2007b. The 1-Laplacian, the ∞-Laplacian and differential games. Pages 245–254 of: *Perspectives in Nonlinear Partial Differential Equations*. Contemp. Math., vol. 446. Amer. Math. Soc., Providence, RI.

Evans, Lawrence C., and Gariepy, Ronald F. 2015. *Measure Theory and Fine Properties of Functions*. Revised ed. Textbooks in Mathematics. CRC Press, Boca Raton, FL.

Evans, L., and Souganidis, P. 1984. Differential games and representation formulas for solutions of Hamilton–Jacobi–Isaacs equations. *Indiana Univ. Math. J.*, **33**(5), 773–797.

Evans, L. C., and Spruck, J. 1991. Motion of level sets by mean curvature. I. *J. Differential Geom.*, **33**(3), 635–681.

Evans, L. C., and Spruck, J. 1992a. Motion of level sets by mean curvature. II. *Trans. Amer. Math. Soc.*, **330**(1), 321–332.

Evans, L. C., and Spruck, J. 1992b. Motion of level sets by mean curvature. III. *J. Geom. Anal.*, **2**(2), 121–150.

Evans, Lawrence C., and Spruck, Joel. 1995. Motion of level sets by mean curvature. IV. *J. Geom. Anal.*, **5**(1), 77–114.

Fabes, E., and Stroock, D. 1984. The L^p-integrability of Green's functions and fundamental solutions for elliptic and parabolic equations. *Duke Math. J.*, **51**(4), 997–1016.

Fleming, W., and Soner, H. 2006. *Controlled Markov Processes and Viscosity Solutions.* Second ed. Stochastic Modelling and Applied Probability, vol. 25. Springer, New York.

Fleming, W., and Souganidis, P. 1988. Two-player, zero-sum stochastic differential games. Pages 151–164 of: *Analyse mathématique et applications.* Gauthier-Villars, Montrouge.

Fleming, W., and Souganidis, P. 1989. On the existence of value functions of two-player, zero-sum stochastic differential games. *Indiana Univ. Math. J.*, **38**(2), 293–314.

Friedman, A. 1971. *Differential Games.* Pure and Applied Mathematics, vol. XXV. Wiley-Interscience [A division of John Wiley & Sons, Inc.], New York-London.

Gilbarg, D., and Trudinger, N. S. 2001. *Elliptic Partial Differential Equations of Second Order.* Classics in Mathematics. Springer-Verlag, Berlin. Reprint of the 1998 edition.

Han, Q., and Lin, F. 2011. *Elliptic Partial Differential Equations.* Second ed. Courant Lecture Notes in Mathematics, vol. 1. Courant Institute of Mathematical Sciences, New York; American Mathematical Society, Providence, RI.

Huaroto, G., Pimentel, E., Rampasso, G., and Świȩch, A. 2020. A fully nonlinear degenerate free transmission problem. *arXiv preprint arXiv:2008.06917.*

Imbert, C. 2011. Alexandroff–Bakelman–Pucci estimate and Harnack inequality for degenerate/singular fully non-linear elliptic equations. *J. Differential Equations*, **250**(3), 1553–1574.

Imbert, C., and Silvestre, L. 2013. $C^{1,\alpha}$ regularity of solutions of some degenerate fully non-linear elliptic equations. *Adv. Math.*, **233**, 196–206.

Imbert, C., and Silvestre, L. 2016. Estimates on elliptic equations that hold only where the gradient is large. *J. Eur. Math. Soc. (JEMS)*, **18**(6), 1321–1338.

Isaacs, R. 1965. *Differential Games. A Mathematical Theory with Applications to Warfare and Pursuit, Control and Optimization.* John Wiley & Sons, Inc., New York-London-Sydney.

Iwaniec, T., and Manfredi, J. 1989. Regularity of p-harmonic functions on the plane. *Rev. Mat. Iberoamericana*, **5**(1–2), 1–19.

Jensen, R. 1988. The maximum principle for viscosity solutions of fully nonlinear second order partial differential equations. *Arch. Rational Mech. Anal.*, **101**(1), 1–27.

Jensen, R., Lions, P.-L., and Souganidis, P. E. 1988. A uniqueness result for viscosity solutions of second order fully nonlinear partial differential equations. *Proc. Amer. Math. Soc.*, **102**(4), 975–978.

Koike, S., and Świȩch, A. 2012. Local maximum principle for L^p-viscosity solutions of fully nonlinear elliptic PDEs with unbounded coefficients. *Commun. Pure Appl. Anal.*, **11**(5), 1897–1910.

Kovats, J. 2009a. Differentiability properties of solutions of nondegenerate Isaacs equations. *Nonlinear Anal.*, **71**(12), e2418–e2426.

Kovats, J. 2009b. Value functions and the Dirichlet problem for Isaacs equation in a smooth domain. *Trans. Amer. Math. Soc.*, **361**(8), 4045–4076.

Kovats, J. 2012. The minmax principle and $W^{2,p}$ regularity for solutions of the simplest Isaacs equations. *Proc. Amer. Math. Soc.*, **140**(8), 2803–2815.

Kovats, J. 2016. On the second order derivatives of solutions of a special Isaacs equation. *Proc. Amer. Math. Soc.*, **144**(4), 1523–1533.

Krylov, N. V. 1982. Boundedly inhomogeneous elliptic and parabolic equations. *Izv. Akad. Nauk SSSR Ser. Mat.*, **46**(3), 487–523, 670.

Krylov, N. V., and Safonov, M. V. 1980. A property of the solutions of parabolic equations with measurable coefficients. *Izv. Akad. Nauk SSSR Ser. Mat.*, **44**(1), 161–175, 239.

Kuusi, T., and Mingione, G. 2012. Universal potential estimates. *J. Funct. Anal.*, **262**(10), 4205–4269.

Kuusi, T., and Mingione, G. 2013. Gradient regularity for nonlinear parabolic equations. *Ann. Sc. Norm. Super. Pisa Cl. Sci. (5)*, **12**(4), 755–822.

Kuusi, T., and Mingione, G. 2014a. Riesz potentials and nonlinear parabolic equations. *Arch. Ration. Mech. Anal.*, **212**(3), 727–780.

Kuusi, T., and Mingione, G. 2014b. The Wolff gradient bound for degenerate parabolic equations. *J. Eur. Math. Soc. (JEMS)*, **16**(4), 835–892.

Ladyzhenskaya, O., and Ural'tseva, N. 1968. *Linear and Quasilinear Elliptic Equations*. Translated from the Russian by Scripta Technica, Inc. Translation editor: Leon Ehrenpreis. Academic Press, New York-London.

Li, D., and Zhang, K. 2015. $W^{2,p}$ interior estimates of fully nonlinear elliptic equations. *Bull. Lond. Math. Soc.*, **47**(2), 301–314.

Lin, F. 1986. Second derivative L^p-estimates for elliptic equations of nondivergent type. *Proc. Amer. Math. Soc.*, **96**(3), 447–451.

Lindgren, E., and Lindqvist, P. 2017. Regularity of the p-Poisson equation in the plane. *J. Anal. Math.*, **132**, 217–228.

Lindqvist, P. 2006. *Notes on the p-Laplace Equation*. Report. University of Jyväskylä Department of Mathematics and Statistics, vol. 102. University of Jyväskylä, Jyväskylä.

Lindqvist, P. 2016. *Notes on the Infinity Laplace Equation*. SpringerBriefs in Mathematics. BCAM Basque Center for Applied Mathematics, Bilbao; Springer, Cham.

Lions, P.-L. 1982. *Generalized Solutions of Hamilton–Jacobi Equations*. Research Notes in Mathematics, vol. 69. Pitman (Advanced Publishing Program), Boston, MA-London.

Lions, P.-L. 1983. A remark on Bony maximum principle. *Proc. Amer. Math. Soc.*, **88**(3), 503–508.

Lions, P.-L., and Souganidis, P. 1988. Viscosity solutions of second-order equations, stochastic control and stochastic differential games. Pages 293–309 of: *Stochastic Differential Systems, Stochastic Control Theory and Applications (Minneapolis, Minn., 1986)*. IMA Vol. Math. Appl., vol. 10. Springer, New York.

Maggi, F. 2012. *Sets of Finite Perimeter and Geometric Variational Problems*. Cambridge Studies in Advanced Mathematics, vol. 135. Cambridge University Press, Cambridge.

Mingione, G. 2011. Gradient potential estimates. *J. Eur. Math. Soc. (JEMS)*, **13**(2), 459–486.

Mooney, C. 2015. Harnack inequality for degenerate and singular elliptic equations with unbounded drift. *J. Differential Equations*, **258**(5), 1577–1591.

Mooney, C. 2019. A proof of the Krylov–Safonov theorem without localization. *Comm. Partial Differential Equations*, **44**(8), 681–690.

Nadirashvili, N., and Vlăduţ, S. 2007. Nonclassical Solutions of Fully Nonlinear Elliptic Equations. *Geom. Funct. Anal.*, **17**(4), 1283–1296.

Nadirashvili, N., and Vlăduţ, S. 2008. Singular viscosity solutions to fully nonlinear elliptic equations. *J. Math. Pures Appl. (9)*, **89**(2), 107–113.

Nadirashvili, N., and Vlăduţ, S. 2011. Singular solutions of Hessian fully nonlinear elliptic equations. *Adv. Math.*, **228**(3), 1718–1741.

Nadirashvili, N., Tkachev, V., and Vlăduţ, S. 2014. *Nonlinear Elliptic Equations and Nonassociative Algebras*. Mathematical Surveys and Monographs, vol. 200. American Mathematical Society, Providence, RI.

Nicolau, A., and Soler i Gibert, O. 2019. Approximation in the Zygmund class. *Journal of the London Mathematical Society*, **101**(1), 226–246.

Niculescu, C., and Persson, L.-E. 2006. *Convex Functions and Their Applications*. CMS Books in Mathematics/Ouvrages de Mathématiques de la SMC, vol. 23. Springer, New York.

Oberman, A., and Silvestre, L. 2011. The Dirichlet problem for the convex envelope. *Trans. Amer. Math. Soc.*, **363**(11), 5871–5886.

Pimentel, E., and Teixeira, E. 2016. Sharp Hessian integrability estimates for nonlinear elliptic equations: an asymptotic approach. *J. Math. Pures Appl.*, **106**(4), 744–767.

Pimentel, E., and Urbano, J. M. 2021. Existence and improved regularity for a nonlinear system with collapsing ellipticity. (English summary.) *Ann. Sc. Norm. Super. Pisa Cl. Sci (5)*, **22**(3), 1385–1400.

Pimentel, E., Rampasso, G., and Santos, M. 2020. Improved regularity for the *p*-Poisson equation. *Nonlinearity*, **33**(6), 3050–3061.

Rockafellar, R. T. 1997. *Convex Analysis*. Princeton Landmarks in Mathematics. Princeton University Press, Princeton, NJ. Reprint of the 1970 original, Princeton Paperbacks.

Savin, O. 2007. Small perturbation solutions for elliptic equations. *Comm. Partial Differential Equations*, **32**(4–6), 557–578.

Silvestre, L., and Sirakov, B. 2014. Boundary regularity for viscosity solutions of fully nonlinear elliptic equations. *Comm. Partial Differential Equations*, **39**(9), 1694–1717.

Silvestre, L., and Teixeira, E. 2015. Regularity estimates for fully non linear elliptic equations which are asymptotically convex. Pages 425–438 of: *Contributions to Nonlinear Elliptic Equations and Systems*. Springer, Cham.

Stein, E. M. 1970. *Singular Integrals and Differentiability Properties of Functions*. Princeton Mathematical Series, No. 30. Princeton University Press, Princeton, NJ.

Święch, A. 1996. Another approach to the existence of value functions of stochastic differential games. *J. Math. Anal. Appl.*, **204**(3), 884–897.

Święch, A. 1997. $W^{1,p}$-interior estimates for solutions of fully nonlinear, uniformly elliptic equations. *Adv. Differential Equations*, **2**(6), 1005–1027.

Teixeira, E. 2014a. Regularity for quasilinear equations on degenerate singular sets. *Math. Ann.*, **358**(1-2), 241–256.

Teixeira, E. 2014b. Universal moduli of continuity for solutions to fully nonlinear elliptic equations. *Arch. Ration. Mech. Anal.*, **211**(3), 911–927.

Teixeira, E. 2016. Geometric regularity estimates for elliptic equations. Pages 185–201 of: *Mathematical Congress of the Americas*. Contemp. Math., vol. 656. Amer. Math. Soc., Providence, RI.

Teixeira, E., and Urbano, J. M. 2014. A geometric tangential approach to sharp regularity for degenerate evolution equations. *Anal. PDE*, **7**(3), 733–744.

Teixeira, E., and Urbano, J. M. 2021. Geometric tangential analysis and sharp regularity for degenerate PDEs. In: *Proceedings of the INdAM Meeting "Harnack Inequalities and Nonlinear Operators" in honour of Prof. E. DiBenedetto*. Springer INdAM Ser. Springer, Cham.

Trudinger, N. S. 1988. Hölder gradient estimates for fully nonlinear elliptic equations. *Proc. Roy. Soc. Edinburgh Sect. A*, **108**(1–2), 57–65.

Trudinger, N. S. 1989. On regularity and existence of viscosity solutions of nonlinear second order, elliptic equations. Pages 939–957 of: *Partial Differential Equations and the Calculus of Variations, Vol. II*. Progr. Nonlinear Differential Equations Appl., vol. 2. Birkhäuser Boston, Boston, MA.

Uhlenbeck, K. 1977. Regularity for a class of non-linear elliptic systems. *Acta Math.*, **138**(3-4), 219–240.

Ural'ceva, N. N. 1968. Degenerate quasilinear elliptic systems. *Zap. Naučn. Sem. Leningrad. Otdel. Mat. Inst. Steklov. (LOMI)*, **7**, 184–222.

Urbano, J. M. 2008. *The Method of Intrinsic Scaling*. Lecture Notes in Mathematics, vol. 1930. Springer-Verlag, Berlin.

Villani, C. 2009. *Optimal Transport*. Grundlehren der Mathematischen Wissenschaften [Fundamental Principles of Mathematical Sciences], vol. 338. Springer-Verlag, Berlin.

Winter, N. 2009. $W^{2,p}$ and $W^{1,p}$-estimates at the boundary for solutions of fully nonlinear, uniformly elliptic equations. *Z. Anal. Anwend.*, **28**(2), 129–164.

Zhikov, V. 2008. Solvability of the three-dimensional thermistor problem. *Tr. Mat. Inst. Steklova*, **261**(Differ. Uravn. i Din. Sist.), 101–114.

Zygmund, A. 2002. *Trigonometric Series. Vol. I, II*. Third ed. Cambridge Mathematical Library. With a foreword by Robert A. Fefferman. Cambridge University Press, Cambridge.

Index

Printed in the United States
by Baker & Taylor Publisher Services